Colorado Warbird
Survivors 2003

Warbird Survivors

Colorado Warbird Survivors 2003

◆

A Handbook on where to find them

Harold A. Skaarup

Writers Club Press
New York Lincoln Shanghai

Colorado Warbird Survivors 2003
A Handbook on where to find them

Writers Club Press
an imprint of iUniverse, Inc.

For information address:
iUniverse, Inc.
2021 Pine Lake Road, Suite 100
Lincoln, NE 68512
www.iuniverse.com

The aircraft described in Colorado Warbird Survivors and the locations listed for them may change from time to time. Although every effort has been made to ensure accuracy up to the time of publication, there are always amendments to be made. Updates to any of the information found in this handbook would be greatly appreciated, and every effort will be made to include them in future editions.

ISBN: 0-595-26223-6

Printed in the United States of America

This book is dedicated to the highly professional men and women of the United States Armed Services and the Canadian Forces of North American Aerospace Defense Command (NORAD), and Northern Command (NORTHCOM). Many of them have flown or serviced the military aircraft described in this handbook. Because of their service, you and I can sleep soundly at night. May it continue to be so.

To control the air, aircraft bring certain characteristics which are not shared by land or sea forces—the ability to carry weapons over long ranges at great speed, the ability to concentrate rapidly large forces over a distant point, the ability to switch targets and to surprise and deceive—in a word, flexibility.[1]

---

1. John Pimlot, *Strategy & Tactics of Air Warfare*, Chartwell Books, Secaucus, New Jersey, 1979, p. 80.

Contents

List of Illustrations

Foreword

North America is resplendent in aviation history, both military and civilian. The sheer size of Canada and the United States dictated an early interest in air defense and profoundly influenced our dependence on air travel. It is no wonder that both nations developed as "air-faring" nations. A large part of the leadership that contributed to that development can be traced to our Air Forces. Indeed, our proud military heritage is embodied in the dedicated individuals who have served and continue to do so—and in the marvellous aircraft they have flown.

The preservation and public display of these aircraft is a labor of love for many, including the editor of this book. If you are an enthusiast of military aviation history, or one with a passing interest who simply wishes to learn more, you will find a wealth of information in these well-researched pages.

Kenneth R. Pennie
Lieutenant-General
Deputy Commander in Chief
North American Aerospace Defense Command

Preface

The Colonel Jack Wilhite had been a fighter pilot for a long time. We were sitting down in one of the meeting rooms at the Wings Over the Rockies Air Museum, and in the space of a few minutes he mentioned what it felt like having to eject from a disabled North American F-100 Super Sabre when it lost one of its control surfaces at high speed. I listened intently as he spoke of flying with Captain Steve Ritchie, the USAF's first ace of the Vietnam War.

McDonnell Douglas F-4C Phantom II

Captain Steve Ritchie, made his first and fifth kills in a Phantom which is currently on display at the USAF Academy in Colorado Springs. He knew and had flown with Colonel Robin Olds, General Chuck Yeager and Bob Hoover, legendary pilots I have only read about. Colonel Wilhite currently flies a Chinese-made MiG-17F/JJ-5, Reg. No. NX905DM, built in China in 1977. He presents himself in his two-seat fighter in airshows under the title "Red Storm." It is the same kind of aircraft the veteran North Vietnamese fighter pilot Colonel Tomb used to shoot down 13 American combat aircraft until Lt. Randy Cunningham shot him down on 10 May 1972. I had to go back and re-read the story of Lt. Cunningham's and Lt. JG Willie Driscoll's engagement with Colonel Tomb after our conversation. Lou Drendel had interviewed the these two combat veterans for his book "And Kill MiGs." Lt. Cunningham told Lou the story something like this:

> "I bored in on the 17...head on. Suddenly, his whole nose lit up like a Christmas Tree! I had forgotten that the A-4's didn't shoot at you, but this guy was really spitting out the 23-mm and 37-mm! I pulled hard, up in the vertical, figuring that the MiG would keep right on going for home. I looked back and...there was the MiG...canopy to canopy with me! He couldn't have been more than thirty feet away...I could see the pilot clearly...leather helmet, goggles, scarf...we were both going straight up, but I was out-zooming him. He fell behind, and as I came over the top, he started shooting. I had given him a predictable flight path and he had taken advantage of it. The tracers were missing me, but not by much! I rolled out, and he pulled in right behind me."
>
> "Now, I don't know if its ego...you know, you don't like to admit that the other guy beat you...or what, but I said: That SOB is really lucky!" Anyway, I told Willie, "Alright, we'll get this guy now!" I pulled down, and I was holding top rudder, trying to knuckle at the nose. As soon as I committed my nose, he pulled right into me! I thought, "Oh, Oh, maybe this guy isn't just lucky after all!" I waited for his nose to commit, then I pulled up into him...that's a rolling scissors. Well here's where my training came into play again. In training, I had fought against Dave Frost in the same situation, and I had learned that if he had his nose too high, I could snap down, using the one G of gravity to advantage, and run out to his six o'clock. I would be a mile, mile and a half out of range before he could get turned around. This is just what happened. We separated, turned around, and engaged again. Same thing. Up into a rolling scissors...advantage...disadvantage...advantage...disadvantage...disadvantage...disengaged, came back, engaged again, and went up into vertical again. This is one of the very few MiG (pilots) that ever fought in the vertical. They like to

fight in the horizontal. We kept engaging, and I could never get enough of an advantage on him to get a shot…everything my airplane did, he reacted to instinctively. He was flying a damn good airplane! Well, we kept at it, with me outzooming him in the vertical, and him shooting every time I got out in front. I thought, "He's going to get lucky one of these times!"

"The next time we started up in the vertical, an idea came to me…I don't know why…your mind just works overtime in a situation like that…anyway, as we're going up, I went to idle and speed brakes…and he shot out in front of me! I think it really surprised him…being out in front for the first time. Anyway, we're both going straight up, and losing speed fast. I was down to 150 knots, and I knew I was going to have to go full burner to hold it. I did, and we both pitched over the top. As he came over, I used rudder to get the airplane to turn to his bellyside. He lost lift coming over the top and, I think, departed the airplane a little bit. I thought. "This is no place to be with a MiG-17…at 150 knots…that slow…he can take it right away from you." But he stayed too long. He was low on fuel, and I think he decided to run. He pitched over the top and started straight down. I went down after him and, though I didn't think the Sidewinder would guide straight down with all the heat of the ground to look at, I squeezed one off anyway. The missile came off the rail and went to his airplane. There was just a little flash, and I thought, "God, it missed him!" I started to fire my last Sidewinder and suddenly…a big flash of flame and black smoke erupted from his airplane. He didn't seem to go out of control, but he flew straight down into the ground. He didn't get out.[1]

This is the kind of imagery and story-telling that came to mind as Colonel Wilhite described flying his own MiG-17. Military aircraft have held a special fascination for me for most of my life. My father served in the RCAF for many years, retiring as a Warrant Officer. As a dependent member of his family, we lived at a number of bases and stations including overseas in Germany and at home in Canada during his service. As both a dependent back then, and in my current service

---

1.   Lou Drendel, "…*And Kill MiGs,*" Squadron Signal Publications, 1974, p. 50.

in the Army, I have had the chance to see NATO Airpower when its list of combat ready aircraft numbered in the thousands. Today, to have hundreds available at any given time would be unusual.

As a member of the Skyhawks Canadian Army Parachute Team, I've participated in a good number of air shows where a handful of former front-line Warbirds may still be seen aloft. Many have been retired to museums, or they are now standing as gate guardians outside their former airbases. These Warbirds are a significant part of our modern history, and they deserve to be remarked upon and remembered. For that reason, I felt there was a need for a handbook that could be used as a guide for aviation enthusiasts to point the way to where you can find those surviving Warbirds in the state of Colorado.

While I was compiling this data, I had the opportunity to speak with many of the pilots and warbird enthusiasts who are still actively flying the handful of original well-restored and airworthy warbirds remaining in Colorado. Joe Thibodeau, for example, who flies both a North American P-51D Mustang named "Crusader," and a Hawker Sea Fury (in Royal Canadian Navy Markings), told me of his experience five years ago intercepting the Commemorative Air Force's Boeing B-29 Superfortress (named Fifi) as it took off from Grand Junction. He had been given a head's up by a friend, and arrived just as "Fifi" was taking off to head for an airshow at Peterson Air Force Base. Joe orbited the B-29 until it was airborne, and then "launched" on the aircraft, joining up alongside the giant bomber in the same way many of his predecessors known as "Little Friends" had done many times during World War II. The closest thing I could relate to with his story, was being onboard our Skyhawk Parachute Exhibition Team's Douglas DC-3/C-47 Dakota transport when one of our local TBM Avenger water-bomber pilots made a "blank-firing" pass at us. Joe and his Mustang flying compadre, Mike Bertz, also had the chance to form up with the B-17 Flying Fortress "Sentimental Journey", and later, on a Heinkel He111 bomber, a warbird which had been widely used by the Germans during the Battle of Britain. It must have been an incredible experience to be flying in the same airspace as these flying legends. It is something many of us who love to attend airshows would like to do. Joe and Mike have their Mustangs on the cover of this book.

There are many others in Colorado like Joe, with incredible aviation stories to tell. I spoke for some time with Walt Barbo, who at the age of 19 flew Lockheed P-38 Lightnings with the 475th Fighter Group in the South Pacific during WWII. He operated out of Ie Shima and in the Indo-China Theater, dropping bombs and napalm in ground support operations. He flew over Hiroshima three days after the Atomic bomb was dropped on it, and remembers seeing smoke still billowing up from the damage that single weapon had done to the city. He later flew his P-38 in Korea. Today he conducts research for the Wings Over the Rockies Air Museum, and helped track down the two FAA Lightnings which had been registered here. One is in pieces in California; the other is on display in Alaska.

Brian Richardson had the privilege of flying the Commemorative Air Force's Consolidated B-24 Liberator "Sentimental Journey" and told me about his hunt for lost warbirds. According to Brian, few people know there are more than 600 military crash sites just in the state of Colorado. He also knew of the B-17 fuselage over the bar in the Greely Armories, and apparently the ball turret is there as well. The original aircraft had been used in the movie 12 O'clock High. Brian had flown a Stearman for awhile, and had run a flight training school. He also airlifted a B-25 Mitchell from its crash site in Alaska using a CH-54 Skycrane, bringing it to Clear AFB. Now he investigates air crashes, which is how he became aware there are so many sites in Colorado. He spoke of Jim McDonald who kept up a continuous hunt for old warbirds until he died of cancer recently. On one expedition Jim found a nearly mint Ryan PT-22 trainer stowed away in a barn up in a remote part of the state. Perhaps there are more discoveries like that one waiting to happen here.

This brings me to explain what this book is about. Many examples of aircraft that saw service with the United States Army Air Corps

(USAAC), the United States Army Air Force (USAAF), the United States Navy (USN), The United States Marine Corps (USMC), and the United States Coast Guard (USCG) have been or are currently being salvaged and preserved in Colorado, particularly where they are of significant historical interest. The Republic P-47N Thunderbolt at Peterson Air Force Base in Colorado Springs, for example, has been taken off its pylon and is currently being restored inside one of the bases original WWII era hangars, by museum volunteers. Parts are still being sought to make the B-18 Bolo at the WORA&SM more authentic, and the staff at the PWAM would dearly love to get hold of a genuine B-24 as the center show piece of their B-24 Memorial Museum.

The purpose of this handbook is to provide an update to the first edition of Colorado Warbird Survivors published in 2001, and a simple checklist of where the surviving military aircraft in Colorado are now, and to provide a photograph of the major types mentioned. This list is also appended with a brief summary of the aircraft presently on display within the state by location, and a bit of the Warbird's history in the US military. Due to space limitations, a selection of only those Warbirds that can be found in Colorado is provided. If you are interested in other aviation books like this one, they can be found at a number   of   online   bookstores: **www.iUniverse.com**; **www.barnesandnoble.com;** and **www.amazon.com**. Look for the Warbird Survivor series.

Aircraft are traded, museums acquire additional displays, and some are lost to accidents and other causes, therefore, this list is only current to the date of publication. If you have additional information you would like to add, please forward an update to me at 2110 Cloverdale Drive, Colorado Springs, CO, 80920, or e-mail me at **h.skaarup@worldnet.att.net**.

It is my sincere hope that the list of Colorado Warbird Survivors will continue to grow as more of them are recovered and restored. Grant that you find this updated handbook useful. Cheers, Harold A. Skaarup

Acknowledgements

I would like to acknowledge and thank each and every member of the museum staffs, particularly the volunteers, of the *Peterson Air & Space Museum*, the *Pueblo Weisbrod Memorial Museum, the Wings Over the Rockies Air Museum,* The *Fort Carson Military Museum*, and the *United States Air Force Academy*, for their patience and assistance in helping me to ensure that the data that has gone into the compilation of this handbook is as complete as it can be to the time of printing. Each and every visitor to your museums owes you that same appreciation, and to all of you, thank you for preserving our aviation heritage.

I would very much like to specifically thank former B-17 pilot Bruce Borman with the WORAM; Walt Barbo, former P-38 pilot with the WORAM; Dusty Rhodes with the WORAM; Lance Barber, who knows the history of the F-100 inside out; Brian Richardson, aviation historian with the WORAM; Don Welsh with the WORAM; Jason B. Unwin with the PWAM; Joseph Thibodeau, current Mustang and Sea Fury pilot and his son J.P. Thibodeau, who fly out of Centennial Airport, particularly for the wonderful cover photo taken by Greg Gedny; and Colonel (retired) Jack Wilhite, former F-100 pilot and present MiG-17 pilot at Jefferson County Airport; and Flint Whitlock with the Colorado Military History Museum.

The list of advisors and people who have given assistance with the research for this handbook include Jason Unwin, who opened up the PWAM archives to help me with the research of serial numbers for the aircraft in the museum, and who pointed out the fact that of the 600 aircraft crashes known to have taken place in Colorado, 360 of them were B-24 Liberators operating out of Pueblo, up to the last one to crash not far from the present day airfield on 25 July 1945; Allen E. Seamans, a former navigator on Consolidated B-24 Liberators with the

8th Air Force, 389th Bomb Group, 564th Bomber Squadron flying over Europe and a volunteer with the PWAM; Rudy Eskra, president of the PWAM and former Boeing B-17 Flying Fortress pilot. Rudy later flew B-29s, C-119s and AT-10s, and now keeps a Stearman biplane in the air; Steve Csogi, who flew B-24s, PT-17s, SNJ/AT-6 Texans and B-25 Mitchells during the war; Katy Heuerman, primary mechanic for the PWAM, and C-172 pilot; and Brother Paul R. Cooper, Electrician and former top gunner in a Martin PBM Mariner, operating with the HEDRON-1 HQ Squadron flying out of San Diego, California in 1945, and now a volunteer at the Pueblo Weisbrod International B-24 Memorial Museum; Dr. Mary Elizabeth Ruwell, curator for the Peterson Air and Space Museum; SSGT Collins; Peg Mesco, who saw the light; Erv Smalley, retired Senior MSgt who worked on F-106 Delta Darts for 20 years; Dave Lankford, a retired radar technician who worked on virtually all NORAD aircraft equipped with Hughes radars; and Michael A. Blaugher for his monumental *"Guides to Aircraft Museums in Canada and the USA."* Their support and assistance in tracking down the information pertaining to each of the individual aircraft listed here was invaluable.

I extend special thanks to the archival staffs in the Library at *the United States Air Force Academy* in Colorado Springs, Colorado; the historians and staff of the *United States Air Force Museum* in Dayton, Ohio; and the *National Museum of Naval Aviation* historians and staffs in Pensacola, Florida. Their patience and assistance has been invaluable in helping me to ensure that the data that has gone into the compilation of this handbook is as complete and accurate as it can be at the time of printing. Their support and assistance in tracking down the information pertaining to each of the individual aircraft listed here is greatly appreciated. Many thanks to each and every one of you.

Blue Skies, Harold A. Skaarup

List of Abbreviations

AFB	Air Force Base
CF	Canadian Forces
CAF	Commemorative Air Force
CO ANG	Colorado Air National Guard
DIA	Denver International Airport
NAS	Naval Air Station
NMNA	*National Museum of Naval Aviation*
NORAD	North American Aerospace Defense Command
PA&SM	*Peterson Air and Space Museum*
PWAM	*Pueblo Weisbrod Air Museum*
USAAC	United States Army Air Corps
USAAF	United States Army Air Force
USAF	United States Air Force
USAFA	United States Air Force Academy
USAFM	*United States Air Force Museum*
USCG	United States Coast Guard
USMC	United States Marine Corps
USN	United States Navy
USNORTHCOM	United States Northern Command
USSPACECOM	United States Space Command
WORA&SM	*Wings Over the Rockies Air and Space Museum*

Introduction

For those of you who are familiar with the airspace over Colorado and its environs, the weather and color of the mountains and the plains that race up to meet them can be incredibly beautiful, particularly when viewed early in the morning from where I work on Cheyenne Mountain.

During my tour of duty with the Canadian Forces detachment based in Colorado Springs, it has been my privilege to serve alongside a good number of the highly professional airmen and women are currently working for NORAD in the Cheyenne Mountain Operations Center, and with the staffs being assigned to the newly created Northern Command, also headquartered in Colorado Springs.

Being in the military has also provided me with the opportunity to visit the state of Colorado, and to examine a number of Colorado's Warbird survivors close up. I have made a specific point of visiting many of the sites where aviation history has been made in the state. These opportunities continue to be a privilege and an honour that is part of my military service, and I would therefore like to share some of the information I gathered with you. It is my hope that this book will show you where to find and view some of Colorado's veteran military aircraft, and to perhaps take an interest in some of the military aviation history that can be found in this warm and friendly state on America's west coast. This book is specifically intended to provide a "where are they" guide for residents and visitors to Colorado who are interested in its rich resources of historical military aircraft.

I have participated in a great number of airshows as both a civilian skydiver and military parachutist. These airshows gave me the opportunity to hear the sound of a P-51 Mustang and a few of its stablemates, as well as the chance to sit in the cockpit of a number of them.

Based on my flight experiences and observations to date, however, I concluded that you should never land in an airplane if you do not want to die in one. (I am equipped with "two perfectly serviceable parachutes" which I pack myself, and you have only one airplane. Also, there is no such thing as a "perfectly serviceable airplane" as any mechanic will tell you).

Although I am in the Army, I've never lost my fascination for old Warbirds. Because of this, I continue to look for, identify and research the locations for as many of them as possible. I then attempt to verify their serial numbers through the *United States Air Force Museum*, and the *National Museum of Naval Aviation*, and to photograph the aircraft wherever they may be on display.

The main reason that I am producing this book is to provide other interested aviation enthusiasts with the kind of guide-book that I would liked to have had before arriving in Colorado to have a look at them. This guidebook should tell the aircraft hunter where he or she may still find these Warbirds and gate guardians, and, where possible, a way to contact the museums and airfields that display them for more information on the aircraft.

I believe that the volunteers who put so much time, effort and energy into maintaining and preserving the numerous retired military and historic aircraft found in Colorado deserve an enormous amount of praise and credit for their work. It is my hope that this handbook, "Colorado Warbird Survivors," provides the information and perhaps an incentive that will bring you to visit their museums and to appreciate the rich resources of aviation heritage they are preserving on your behalf.

Harold A. Skaarup

Aircraft Museums and displays in Colorado

Aspen, CO.

Aero L-39 Albatros (Serial No. 332801), Reg. No. N43882. J & E Holdings, PO Box 12345, Aspen, CO 81612-9237.
Beechcraft T-34A Mentor (Serial No. G-285), Reg. No. N874Z. William S. Greenwood, Box 4778, Aspen, CO 81612.
Supermarine Spitfire Mk. IX (Serial No. TE308), Reg. No. N308WK, William S. Greenwood, Box 4778, Aspen, CO 81612. This aircraft is a two-seat variant, with previous service with the Irish Air Force.

Aurora, CO.

Folland Fo.141 Gnat T. Mk. 1 (Serial No. FL592) Reg. No. N998XR. John Robert Mulvey, Astre Air International, 6224 S. Jericho Ct., Aurora, CO 80016-1297.
Folland Fo.141 Gnat T. Mk. 1 (Serial No. FL504) Reg. No. N694XM. John Robert Mulvey, Astre Air International, 6224 S. Jericho Ct., Aurora, CO 80016-1297.
Hunting P66 Pembroke (Serial No. PAC/W/781), Reg. No. N4273C. John Robert Mulvey, Astre Air International, 6224 S. Jericho Ct., Aurora, CO 80016-1297.
North American AT-6A Texan (Serial No. 78-7375), Reg. No. N90629, Bob Ford, 12091 East Canal Drive, Aurora, CA 80011.
Piper J-3C-50S Cub (Serial No. 18520), Reg. No. N98349. Steven J. Beckman, 5763 S Orlenas St., Aurora, CO 80015-3501.

Aurora, CO. **Buckley Air Force Base**, STOP # 24, 140 TFG, 80011-9599. Tel: 303-366-5363. Buckley AFB, CO. Internet: **www.cobuck.ang.af.mil/**.

Located in Aurora, a suburb of Denver eight miles East of the city. The former Buckley Air National Guard Base has become an active duty facility and is home to the 821st Space Group, 140th Wing, which flies F-16 Fighting Falcon fighters. Other units on base include HQ Colorado ANG; 227th Air Traffic Control Flight (ANG); 240th Civil Engineering Flight (ANG); Air Force Office of Special Investigations, Detachment 402; the Colorado Army National Guard's Aviation Support Facility; Navy Reserve; Marine Corps Reserve; ARNG; and Air Force Units. The ANGB was activated on 1 April 1942 as a gunnery training facility. The ANG assumed control from the US Navy in 1959.

The base was named after 1st Lt. John Harold Buckley, a World War I flier from Longmount, Colorado. Buckley lost his life in France, when his aircraft was shot down 17 September 1918, behind German lines.

Buckley began as an extension of Lowry Army Air Field, Colo., to train World War II troops. In 1947, the Navy took over command and renamed it Naval Air Station Denver. When the Guard became host in 1960, the installation changed to Buckley Air National Guard Base, making it the first Air National Guard base in the nation."

The base was re-designated an Air Force Base on 2 October 2000. A number of helicopters formerly on display at Buckley have been moved, possibly to Greely.

Bell UH-1 Iroquois (Serial No. 65-09840)
Bell UH-1 Iroquois (Serial No. 65-09874)
General Dynamics F-16 Fighting Falcon (USAF Serial No. 79-373). In front of the Base HQ building.
North American AT-6D Texan (Serial No. 61961). On the F-16 Flight Line.

North American F-86D Sabre (Serial No. 51-13010) (painted as 51-2884). In front of the Base HQ building. Silver with "Minute Man" colors.
North American F-100A Super Sabre (Serial No. 53-1578), painted as 54-1897. In front of the Base HQ building. Camouflage colors.
North American F-100A Super Sabre (Serial No. 63299). In front of the Base HQ building. Silver.
Sikorsky SH-34J/S-58 Choctaw (Serial No. 50213). In front of the Navy and Marine HQ building.
Vought A-7D Corsair II (Serial No. 70-001). In front of the Base HQ building.
Vought A-7D Corsair II (Serial No. 69-6230). In front of the Navy and Marine HQ building. USS Constellation colors.

Three aircraft previously at Buckley have been moved:

North American F-100D Super Sabre (Serial No. 56-3299)
Vought A-7D Corsair II (Serial No. 69-6242)
Vought A-7D Corsair II (Serial No. 73-0996)

Basalt, CO.

De Havilland DHC-2 Beaver Mk. I (Serial No. 151), Reg. No. N5161, Wayne G. Rudd, 132 Park Ave., Basalt, CO 81621.
Grumman TBM-3E Avenger (Serial No. 5632), Reg. No. N81865. Wayne G. Rudd, 132 Park Ave., Basalt, CO 81621.
North American P-51D Mustang (Serial No. 122-51619), Reg. No. N3333E. Wayne G. Rudd, 132 Park Ave., Basalt, CO 81621.
North American T-28A Trojan (Serial No. 51-7606), Reg. No. N9103F. Wayne G. Rudd, 132 Park Ave., Basalt, CO 81621.

Berthoud, CO.

Beechcraft T-34A Mentor (Serial No. G-735), Reg. No. N3145C. Roger L. Gibbons, 608 W County Road, 2E, Berthoud, CO 80513.

Boulder, CO.

Bell UH-1H Iroquois (Serial No. 71-20310), Reg. No. N3140U. Pan American Partners, 312 Hollyberry Lane, Boulder, CO 80303-5231.
Grumman G-44 Widgeon (Serial No. 1235), Reg. No. N9933H. Windy Hill Widgeon Works, c/o Marilyn M. Soby, 9998 Hold Hill Road, Boulder, CO 80302.
Grumman HU-16C Albatross (Serial No. 17162), Reg. No. N51025, Howard W. Selby III, 244 Arroyo Chico Salena Star Rte., Boulder, CO.
North American T-28A Trojan (Serial No. 51-3616), Reg. No. N28TW. 844th Squadron Inc., 2871 Loma Place, Boulder, CO 80301.

Breckenridge, CO.

North American T-28C Trojan (Serial No. 146283), Reg. No. N1184T. Eagle Air II, PO Box 322, Breckenridge, 804243322.CO

Brighton, CO.

Beechcraft T-34B Mentor (Serial No. 140819), Reg. No. N434D. Delta Enterprises Leasing, 15871 Duquesne Circle, Brighton, CO 80601-3856.
North American AT-6C Texan (Serial No. 88-13651), Reg. No. N696RE, Russell E. Emicle, 15885 Delta Ct., Brighton, CO 8061-3863.

Broomfield, CO. Jefferson County Airport.

Beech Model 17 Traveler Staggerwing (Serial No.)
De Havilland D.H. 112 Venom (Serial No. 737), Reg. No. N21MJ. M & M Aircraft Sales Coy, 11740 Airport Way, Hangar 36F, Broomfield, CO 80021-2558. Swiss Air Force.

Folland Fo.141 Gnat T. Mk. 1 (Serial No. FL504) Reg. No. N694XM. Michael W. Bertz, 11740 Airport Way 35-C, Broomfield, CO 80021

Folland Fo.141 Gnat T. Mk. 1 (Serial No. FL562) Reg. No. N572XR. Michael W. Bertz, 11740 Airport Way 35-C, Broomfield, CO 80021.

Hunting Percival P.84 Jet Provost T-3 (Serial No.) Michael W. Bertz, 11740 Airport Way 35-C, Broomfield, CO 80021.

Mikoyan Gurevich MiG-17F/JJ-5 (Serial No. 551604), Reg. No. N905DM. Built in China in 1977, this two-seater is armed with one 23-mm cannon. Colonel (USAF Ret'd) Jack E. Wilhite, Red Storm Airshows, 6400 Crestbrook Drive, Morrison, CO.

North American T-6G Texan (Serial No. 52-8236-1), Reg. No. N4269R. Jack Thunder, PO Box 274, Broomfield, CO 80020.

North American P-51D Mustang (Serial No. 45-11636N), Reg. No. N11636N. Michael W. Bertz, 11740 Airport Way 35-C, Broomfield, CO 80021.

Vought A-7 Corsair II (Serial No.), parts

Broomfield, CO. Jeffco Warbird Group

The Jeffco Warbird Group was formed by Michael Bertz, Col., CO ANG, to combine the efforts of several groups of Warbird owners and operators who are located at Jefferson County Airport (Jeffco/KBJC), Broomfield, Colorado. The mission of the group is to preserve, maintain and fly historic Military propeller and jet powered aircraft.

The Jeffco Warbird Group has participated in many airshows including several Cheyenne Frontier Days Airshows, The Wyoming ANG 40th Anniversary Open House, Airshow Colorado 1997, and the Ellsworth AFB 50th Anniversary Airshow and Open House.

Many historic aircraft participate with the Jeffco Warbird Group including a rare North American P-51D Mustang. Jet aircraft which participate in the Jeffco Warbird Group include a Folland Gnat, DeHavilland Venom, MiG-17 and a Jet Provost T Mk. 5.

The Jeffco Warbird Group is available for Air Shows and Open Houses. For information about Jeffco Warbirds, call (303) 460-1156. *(Internet:* **http://www.warbirdrecovery.com/jeffco.html**). Tel: 303-460-1156. Fax: 303-464-7576. E-mail: **info@warbirdrecovery.com**

Broomfield, CO. Warbird Recovery

Warbird Recovery is committed to the International recovery and restoration of rare and historic military aircraft. The first aircraft recovery trip took place in the Summer of 1993 in Siberian Russia. The mission was to recover rare Japanese and US fighters. However, that mission is still ongoing and has not been completed. Aircraft which have been successfully recovered and brought to the United States by Warbird Recovery include several Messerschmitt Bf-109-E fighters and a British Hawker Hurricane.

Warbird Recovery is always looking for additions to the collection and would welcome any information you might have on locations of military aircraft or parts that might be of interest. Tel: 303-460-1156. Fax: (303) 464-7576.

E-mail: **info@warbirdrecovery.com**.

Messerschmitt Bf 109F-4. This aircraft was built in 1942 in Leipzig, Germany. Records show that this aircraft was a Norwegian patrol aircraft which flew the Murmansk sector of Russia and was shot down in World War II by Russian forces. It remained where it came to rest in Northern Russia until it was recovered by Warbird Recovery in the winter of 1994. All remaining pieces were shipped to the Warbird Recovery facility in Broomfield, Colorado.

A Messerschmitt Bf 109E-7 is also under restoration at Warbird Recovery. The Messerschmitt Bf 109 aircraft was so successful that when production of the Bf 109 ceased in 1956, over 35,000 had been built. The performance of the Bf 109 was equal to the Spitfire and superior to the Hurricane. There is now only one flying original Bf 109 in the world. Warbird Recovery has begun restoring this aircraft to original flying condition.

Hawker Hurricane MKI RAF Serial No. P3311. This aircraft was built in 1940 by Hawker Aircraft Limited, England. Records show that this aircraft took off from North Weald Aerodrome on 27 May 1940 at 1550-hrs. The pilot was Flt. Lt. Lee who engaged with German He 111's over Dunkirk, France. Flt Lt. Lee was shot down over the English Channel. After bailing out he was rescued at sea. It is uncertain if he was shot down by an He 111 or by an Me 110 bomber.

The remains of the Hurricane were recovered off the beach at Dunkirk and were moved to the London, England area. In 1995 all remaining pieces were shipped to Fort. Collins, CO by Warbird Recovery. Warbird Recovery has begun restoring this aircraft to original flying condition.

Broomfield, CO. Spirit of Flight Air Museum

The Spirit of Flight Air Museum was founded in August 1998 by Gordon R. Page. The Spirit of Flight Museum is dedicated to the acquisition, restoration, preservation and display of rare and historical Military Aircraft and related artifacts. The goal of the Museum is to display significant Military Aircraft at a new facility located at Jefferson County Airport in Broomfield, Colorado and to educate the general public about the history of the world events which these aircraft participated in. The Museum will also hold frequent flying events to display it's aircraft to the General Public. Along with the aircraft, the Museum facility will also house an aviation library, aviation art gallery, meeting areas, restoration area, gift shop, snack area and a special area which will honor individuals and companies which have made significant contributions to Aviation in the Denver and Rocky Mountain region. E-mail: **spiritofflight@warbirdrecovery.com**. *(Internet:* **http:// www.warbirdrecovery.com/sof.html)**.

Canon, CO. Fremont County Airport

McDonnell F-4C Phantom (Serial No. 63-7551), mounted on a pylon

Lockheed T-33A Shooting Star (Serial No. 57-0575). This aircraft was originally on display at Canon; it is presently unlocated.

Carr, CO.

North American AT-6F Texan (Serial No. 44-81976), Reg. No. N7464C, Kenneth J. Burham, 6932 County Road 110, Carr, CO 80612-8908.

Carbondale, CO.

North American AT-6F Texan (Serial No. 44-81857), Reg. No. N81854, Wayne Rudd, 52 Flying Fish Road, Carbondale, CO 81623-9566.

Colorado Springs, CO.

Bell OH-58C Kiowa (Serial No. 70-15136), Reg. No. N196PD. City of Colorado Springs Police Department, 705 S. Nevada Ave., Colorado Springs, CO 80903.
Bell OH-58C Kiowa (Serial No. 68-16947), Reg. No. N198PD. City of Colorado Springs Police Department, 705 S. Nevada Ave., Colorado Springs, CO 80903.
Boeing Model 367 KC-97L Stratocruiser (Serial No. 53-283), Texas Air National Guard, Steve Kanatzar, Solo's Restaurant, 1665N Newport Road (off Fountain, one block east of Powers). Tel: 570-7656. Hours: Sun.-Thurs., 11 a.m.-9 p.m.; Fri.-Sat., 11 a.m.-10 p.m.
Consolidated Vultee BT-13A Valiant (Serial No. 7333), Reg. No. N67232. William R. Sutherland, 226 W Platt, Colorado Springs, CO 80900.

Hiller UH-12B Raven (Serial No. JR-1), Reg.No. N5104V.

Hiller UH-12D Raven (Serial No. 2118), Reg. No. N9740C. Mark A. Henry, 5015 Goodnight Ct, Colorado Springs, CO 80922.

Hiller OH-23D (Serial No. 58-5454), Reg. No. N604RA. Trans Aero Ltd., 715 S. Circle Drive, Suite 102, Colorado Springs, CO 80910.

Hunting Percival P.84 Jet Provost T-3 (Serial No. PACW 6328), Reg. No. N4427Q. Charles B. Simpson, 1330 Oak Hills Drive, Colorado Springs, CO 80919.

North American AT-6C Texan (Serial No. 27352), Reg. No. N7058C, Mrs. J. A. Flesher, 6150 Daylight Point, Colorado Springs, CO 80918-7678. Possibly sold to a buyer in Parkland Estates.

North American AT-6D Texan (Serial No. 90672), Reg. No. N6643C, Richard D. Janitell, 325 Haversham Drive, Colorado Springs, CO 80906-7204.

North American AT-6D Texan (Serial No. 88-14945), Reg. No. N800SH, Barton Aviation, 15 Friendship Lane, Colorado Springs, CO 80904-1810. This aircraft formerly served with the South African Air Force.

North American T-28C Trojan (Serial No. 140520), Reg. No. N678MC. Michael B. Cranford, 4985 Iron Horse Trail, Colorado Springs, CO 80917.

Peterson AFB, CO. Internet: **www.peterson.af.mil/**. Located five miles east of Colorado Springs, Peterson AFB houses the 21st Space Wing, North American Aerospace Defense Command, US Northern Command, 302nd Airlift Wing (Reserve) and provides support functions for Cheyenne Mountain Air Force Station and Schriever Air Force Base.

Colorado Springs, CO. **Edward J. Peterson Air & Space Museum**, 21st Space Wing/MU, 150 East Ent Avenue, Peterson Air Force Base, CO, 80914-1303. Curator: Dr. Mary Elizabeth Ruell. Telephone: 719-556-4915. FAX: 719-556-4916. Chairman, Board of Directors: Major General George Douglas, USAF (Retired).

The museum is located on Peterson Air Force Base in Building 981 at the intersection of Peterson and Ent Boulevards. Peterson Air Force Base is approximately seven miles east of Colorado Springs on Highway 24. Hours of Operation: The museum is open from 08:30 AM until 4:30 PM Tuesday through Saturday, but closed Sunday, Monday and Federal holidays. Admission is free.

The Museum is housed in the original Art Deco style Colorado Springs Municipal Airport Passenger Terminal building constructed in 1941. Located within an 8.3-acre Colorado State Historic District, the Museum preserves and portrays the rich aviation and space history of Colorado Springs and Peterson Air Force Base. Exhibits feature the WW II Peterson Army Air Base, the Bi-national (United States and Canadian) North American Aerospace Defense Command, US Air Defense Command and the US Air Force Space Command. The Museum's Airpark collection includes sixteen aircraft and four missiles.

The museum offers unguided "discovery" tours and is fully handicapped accessible. A Braille guidebook is available for the visually impaired. Non-military visitors may be required to show a current driver's license, automobile registration and proof of insurance to the Gate Guard or Base Visitor's Center located at the Main Gate to gain entry to Peterson Air Force Base to visit the Museum). A temporary pass will then be issued to visit the Museum). Internet: **http:// www.petemuseum.org**.

Peterson AFB is located at the eastern edge of Colorado Springs. It is host to the 21st Space Wing, and its major tenants are Northern Command (NORTHCOM), North American Aerospace Command (NORAD); Air Force Space Command; Army Space Command; 302d Airlift Wing (AFRC); and the Edward J. Peterson Air and Space Museum). The AFB was first activated in 1942. It is named for 1st Lt. Edward J. Peterson, who was killed in the crash of his Lockheed P-38 Lightning on the 8th of August 1942.

The following aircraft and missiles are on static display at the PA&SM:

Avro CF-100 Mk. 5C Canuck (Serial No. 100779)
Boeing CIM-10A BOMARC Missile (Serial No. 59-2051)
Convair F-102A Delta Dagger (Serial No. 56-1109)
Convair GF-106A Delta Dart (Serial No. 59-134)
Curtis P-40E Warhawk (Serial No. 1356255), mounted on a pylon.
Lockheed T-33A Shooting Star (Serial No. 57-575)
Lockheed F-94C Starfire (Serial No. 50-1006)
Lockheed F-104C Starfighter (Serial No. 56-936), mounted on a pylon.
Lockheed EC-121T Warning Star (Serial No. 52-3425)
Martin EB-57E Canberra (Serial No. 55-4279)
McDonnell CF-101B Voodoo (Serial No. 57-381)
McDonnell F-101B Voodoo (Serial No. 58-274), mounted on a pylon.
McDonnell F-15A Eagle (Serial No. 76-024)
McDonnell-Douglas F-4C Phantom (Serial No. 64-0799)
North American F-86L Sabre Dog (Serial No. 53-0782)
Northrop F-89J Scorpion (Serial No. 52-1941)
Republic P-47N Thunderbolt (Serial No. 44-89425). Inside the museum hangar, currently being restored.
U.S. Army Hawk Missile
U.S. Army Nike Ajax Missile
U.S. Army Nike Hercules Missile

Colorado Springs, CO. **United States Air Force Academy**, CO 80840-5151. Tel: 719-333-2025 (Visitor Center), or 333-4674 (Library Special Questions Desk. Internet: **www.usafa.af.mil**. Located in Colorado Springs, the Academy houses the 34th Training Wing; Dean of Faculty, USAF Preparatory School, 10th Air Base Wing and 557th Flying Training Squadron.

"In 1949 the Secretary of Defense, James Forestall, created a committee to recommend a general system of education for service officers. Within a year, this advisory group reported that only an air academy, separate from both West Point and Annapolis, would meet the needs of the air force. When Congress authorized creation of the institution in 1954, the Secretary of the Air Force launched the site selection process. Three "semi-finalists" emerged from the screening: Alton, Illinois; Lake Geneva, Wisconsin, and Colorado Springs." Colorado won. The cadet wing moved into the academy from its temporary campus at Lowry Field in late August 1958. The following June the first class was graduated from the permanent campus."[1]

NORAD has also been in Colorado Springs (first at Ent Air Base, later at Peterson Air Force Base and Cheyenne Mountain) since 1957. Historic aircraft have been on display at the USAFA from its earliest days.

The following aircraft and missiles are on outdoor static display at the United States Air Force Academy:

Boeing B-52D Stratofortress (Serial No. 55-083), black and camouflage colors.
Cessna T-41 Mescalero (Serial No.), Reg. No. N557AF, Academy Airport.
Fairchild A-10 Thunderbolt (Serial No.) During Operation Desert Storm, Capt. Swain was the 1st pilot in history to have an air-to-air shoot down of an enemy helicopter. This aircraft was mounted on a pylon at the USAF Academy in May 2002. Thunderbird Lookout.
General Dynamics GF-16A Fighting Falcon (Serial No. 75-0748), below the Chapel.
Lockheed F-104A Starfighter (Serial No. 55-2967), Academy Airport.

---

1.   Carl Ubbelohde, Maxine Benson and Duane A. Smith, *A Colorado History, Seventh Edition*, Pruett Publishing Company, Boulder, Colorado, 1999, p. 334.

Martin Marietta X-24A/SV5-J "Swallow" Lifting Body (Serial No.), on display beside the Aeronautics Laboratory.

McDonnell-Douglas F-4C Phantom II (Serial No. 66-7463), below the Chapel.

McDonnell-Douglas F-15A Eagle (Serial No. 76-042), below the Chapel.

North American F-86F Sabre (Serial No.52-4749), painted as 24978, was on display at the Academy Airport until August 2002, when it was returned to the USAF Museum.

North American T-38A Talon (Serial No. 59-1602), Thunderbird markings. Thunderbird Lookout.

North American F-100D Super Sabre (55-3503), not located.

North American F-100F Super Sabre (Serial No. 56-3730) "Spirit of St. Louis II", near the BX.

Republic F-105D Thunderchief (Serial No. 60-482), below the Chapel.

Minute Man II Missile, near the Sports Center.

Schweitzer Glider (Serial No.), inside the Visitor's Center.

Colorado Springs, CO. **Butts Airfield**, **Fort Carson**, CO

Bell UH-1 Huey (Serial No. 65-09899), 4th Sqn, 3d Armd Cavalry Regt, Butts Army Airfield.

Bell UH-1 Huey (2 hulks), 10th Special Forces Airborne Group.

Bell AH-1F Cobra Helicopter (Serial No. 023247, HQ 4th Sqn 3rd Armd Cavalry Regt, Building 2350, Titus Street & Barkeley Street.

Bell AH-1F Cobra Helicopter (Serial No. 015687), Main Gate.

Bell OH-58C Kiowa (Serial No. 20403), Butts Army Airfield, Bldg 9658

Sikorsky H-34 Choctaw (Serial No. 53-4547), Third Cavalry Museum).

Craig, CO.

Hiller UH-12D Raven (Serial No. 2351), Reg. No. N7173S. Craig Helicopters Inc., PO Box 1212, Craig, CO 81626-1212.

Del Norte, CO.

Hawker Sea Fury T Mk. 20 (Serial No. 37537), Reg. No. N35SF. Vernon C. McCallister, 57 First Ave., Del Norte, CO 81132. Project.

Denver, CO

Beechcraft T-34 Mentor (Serial No.), Sky Fighters Inc., Centennial Airport.
Beechcraft T-34 Mentor (Serial No.), Sky Fighters Inc., Centennial Airport.
Hawker Sea Fury (Serial No. 37511), Reg. No. NX2545F, FB60, RCN colors. Joseph H. Thibodeau, 155 South Madison Street, Suite 209, Denver, CO 80209. Iraqi wings.
Lockheed 1049C Constellation (Serial No. 4535), Reg. No. N6227C, Proimex International Ltd., 8000 E Girard 400, Denver, CO 80231.
North American AT-6 Texan (Serial No. 7695), Reg. No. N98FP, William M. Reed, 360 Jasmine, Denver, CO 80220.
North American AT-6G Texan (Serial No. 49-3190), Reg. No. N884TA, Steven Cowell, PO Box 370022, Denver, CO 80237-0022.
North American P-51D Mustang (Serial No. 44-74582), Reg. No. N6329T. Joseph H. Thibodeau, 155 South Madison Street, Suite 209, Denver, CO 80209.
North American T-28 Trojan (Serial No.)
Sikorsky S-58 Seabat (Serial No. 148021), Reg. No. N51803. Richard J. Salter, 5067 W. Moncrieff Place, Denver, CO 80212.

The following warbirds have been listed in the FAA Registry for Colorado, but are no longer in the state:

Douglas AD-4DW Skyraider (Serial No. 126882-SF85), Reg. No. N91945. Now in the Marine Aviation Museum, 3511 Silverside Road, Ste 105, 19810-4902.
Grumman OV-1D Mohawk (Serial No. 62-5889), Reg. No. N134GM, Double Eagle Co., 3200 Cherry Creek South Drive, Ste 400, Denver, CO 80209-3265. Possibly sold.
Lockheed P-38L (F-5G) Lightning (Serial No. 42-13400), Reg. No. N55929. American Veterans Memorial Museum, PO Box 10154, Denver, CO 80210. This aircraft crashed in Alaska in 1945, and was recently recovered and restored. It is on display at Elmendorf AFB, Alaska. (Apparently it was registered to Colorado, because the pilot originally came from Denver)
Lockheed P-38L-5 Lightning (Serial No. 44-26961), Reg. No. N6961. This aircraft was owned by John G. Deahl, 3130 East Ohio Way, Denver, CO 80209. He died in a crash with this aircraft near Salt Lake City, Utah. Although the remains were badly burned, the parts are stored with an aviation group in California.

Denver, CO. **Commemorative Air Force**, Denver, CO. The Mile High Wing of the Commemorative Air Force (CAF) received its charter in February 1999. The Mile High Wing had the Lodestar described below. It is now in Texas.

Lockheed Model 18/C-60A-5 Lodestar, (Serial No.) The Army Air Corps accepted this aircraft on 6 October 1943 as a C-60A-5. It was soon delivered to the Royal Canadian Air Force where it saw service during the war. After the war, the Howard 250 executive conversion was done.

Denver, CO. **Colorado Army National Guard Base, Denver,** CO. 69th Bn.

Grumman OV-1D Mohawk (Serial No.), presently unlocated.

Denver, CO. Colorado Military History Museum

The Colorado Military History Museum will, when completed, be a world-class military museum honoring all branches of service and all conflicts in which the U.S. has been involved—from 1775 to the current war on terrorism. The museum is accepting monetary donations as well as military-related artifacts from veterans and their families. Contact: CMHM, PO Box 201714, Denver, CO 80220. Phone: 303-377-4048. E-mail: **flintlock@qwest.net**. Website: **www. coloradomilitaryhistory.org**.

Denver, CO. Denver International Airport (DIA)

Curtiss JN-4D Jenny (Serial No. SC1918 #65) . Hanging in Main Concourse.
Alexander Eaglerock Model A-14 (Serial No. NC205Y). Hanging at opposite end of the Concourse.

Denver, CO. J.W. Duff Aircraft Salvage

Mr. Duff has been salvaging or collecting parts and pieces of old warbirds since 1949, and now has portions of over 500 aircraft, some complete, but mostly fuselages and wings. Mr. Duff has sent several aircraft to the USAF museum, including a Beech U-21 Ute and a Consolidated L-13 Vultee. Duff Aircraft Salvage has parts available for a number of classic warbirds. With over 500 damaged planes and a warehouse and hanger full of parts, J.W. Duff Aircraft Co. in sight of what was Stapleton International Airport's old North-South runway, is one of the largest aircraft "salvage" yards in America. J.W. Duff and his crew retrieve damaged planes for insurance companies and airlines, buys them by a bidding process, and then sells the used parts at a substantial savings to the customer. The Company has a two-level warehouse and over fifteen acres that contain fuselages, wings, surfaces, props, engines, gears, wheels, tires, brakes, and other misc. parts.

None of the aircraft are on display, but inquiries may be directed to: J.W. Aircraft Salvage, 8131 E. 40th Ave., Denver, CO, 80207; or to J.W. Duff Aircraft Company, 3696 Quebec Street, Denver, CO 80200. Tel: 303-399-6010. Fax: 303-321-6790. Open Monday to Friday. Internet: **http://www.jwduffaircraft.com/index.html**.

Beech C-45 Expeditor (Serial No. 43-33283A), Reg. No. N5973V, Beechcraft T-34 Mentor (Serial No. 144098), Reg. No. N5543.
Beechcraft T-34 Mentor (Serial No. 144098), Reg. No. N5543.
Bell 47G Sioux (Serial No. 1528), Reg. No. N2814B.
Bell UH-1H Iroquois (Serial No. 68-15774), Reg. No. N60638.
Bell UH-1H Iroquois (Serial No. 70-16232), Reg. No. N60901.
Bell UH-1L Iroquois (Serial No. 154944), Reg. No. N138US.
Consolidated Vultee/HOPS L-13A Scorpion (Serial No. 47-323), Reg. No. N68845.
De Havilland DHC-2 Beaver L-20A (Serial No. 593), Reg. No. N32414.
De Havilland DHC-2 Beaver U-6A (Serial No. 1919), Reg. No. N90646.
Hiller H-23D Raven (Serial No. 61-3190), Reg. No. N38007.
Hiller H-23D Raven (Serial No. 61-3191), Reg. No. N38008.
North American SNJ-4 Texan (Serial No. 88-10271), Reg. No. N7093C.
Piper L-18C Super Cub (Serial No. 18-2119), Reg. No. N4075A.
Piper L-21A Super Cub (Serial No. 18-919), Reg. No. N6200A.

Denver, CO. **Forney Museum of Transportation**, 4303 Brighton Boulevard
Denver, CO 80216. Tel: 303-297-1113. Internet: **http://www. forneymuseum.com/basicinfo.htm**. E-mail: **forney@frii.net**.

Beech Model 18 (Serial No.), in storage
Rutan Quickie (Serial No.), in storage

Denver, CO. **The Wings Over the Rockies Air and Space Museum**, Hangar Number 1, Building 401, Lowry Campus, 7711 East Academy Blvd, Denver, Colorado, 80220. Tel: 303-360-5360; FAX: 303-5328.

The museum is located at the site of the former Lowry AFB, 7711 East Academy Blvd, Denver, Colorado. The museum is open weekdays from 10:00 AM until 4:00 PM, on Saturdays from 10:00 AM until 2:00 PM, and on Sundays from 1:00 PM until 4:00 PM. Internet: **www.dimensional.com/%7eworm/lobby.html.**
E-mail:**worm@dimensional.com**.

The Wings Over the Rockies Air and Space Museum (WORA&SM) is one of Denver's major historical and educational facilities. The museum is dedicated to preserving the rich history of flight in the Rocky Mountain region, and to teaching the younger generation about the science of aviation and space travel. With 30 aircraft and special exhibits, there is something of interest in the museum for everyone.

The WORA&SM was opened on 1 December 1994. It is Denver's first Aerospace Museum, and is the Official Air and Space Museum of the State of Colorado, as designated by the Colorado State Legislature. The museum is dedicated to education, science, mathematics, and technology. 36 military aircraft, from a B-1A Lancer, back to a vintage B-18A Bolo, and a number of various home-builts, sail planes, restorations, and reproductions are the centerpiece for this educational venue. With the exception of the museum's B-52B, all aircraft are displayed indoors in a climate-controlled environment. The B-52 has a place of honor at the front door to the Museum, and greets all visitors.

In addition to the aircraft on display, the museum is home to other historically significant displays, including the Lafayette Foundation collection of World War I uniforms and memorabilia (for more details, see Appendix C). The Jill Haltom Smith, "Dressed for Duty" collection of Women in Uniform also has a prominent place in the museum.

With 165,000 square feet under one roof, the museum also serves the Denver Metro community as its third largest event venue. The museum plays host to many large events each year, including the Rocky Mountain Air Fair. The annual Air Fair is the largest general aviation exposition in the Rocky Mountain Region, and is normally held during March of each year. Many other local, state, regional, as well as national and international organizations find the museum to be an interesting and unique site for their special event.

Though the WORA&SM is one of the newest air museums in the nation, its rich heritage can be traced back much earlier. The museum is located in historic Hangar One at the former Lowry Air Force Base. Constructed in August 1939, Hangar One was one of the first buildings erected on the fledgling Army airfield. The first unpaved runway became operational at Lowry Field on 4 April 1938. Six of the Air Corps' new B-18A medium bombers were the first aircraft to fly into Lowry, making the 30 block flight from Denver Municipal Airport to Lowry on 13 December 1939, following completion of the new north-south runway.

Throughout Lowry's history, all aircraft assigned and flight operations were in support of the training mission. Bombardiers, Aerial Photographers, Aerial Gunners, as well as supporting personnel trained on B-17s, B-24s and B-29s as well as C-47s at Lowry. Training for maintenance on the Top Secret Norden Bomb Sight was also conducted at Lowry.

The Museum at Lowry started as a camera collection at the Aerial Photographers School in 1967. The size and scope of exhibits in the museum grew, eventually being incorporated as the "Lowry Heritage Foundation" in 1982. The Museum received its first aircraft from the Air Force Museum in 1984, and placed this and all future aircraft on outdoor display.

Transporting these aircraft to Lowry AFB was considerably more difficult than expected. All of Lowry's flying operations had been transferred to the nearby Buckley Air National Guard Base, located six miles to the east, on 30 June 1966. The training mission at Lowry continued however, with all new display and training aircraft being flown into

Buckley, then disassembled and trucked to Lowry. Non-flying aircraft were brought in from Davis Monthan AFB by truck and reassembled at Lowry.

Lowry was placed on the base closure list in 1991, and on 30 September 1994, the American flag came down for the last time at Lowry Air Force Base. With the base closed, the fate of the Heritage Museum was also sealed. Representatives from the Air Force Museum at Wright Patterson AFB immediately began an extensive inventory, packing, and shipping project. This project resulted in all of the significant exhibits, including the extensive camera collection, being shipped to Wright Patterson for storage or display. The display aircraft were listed as available to other Air Force facilities and museums.

A few far-sighted individuals rapidly started a movement to establish the Wings Over the Rockies Air and Space Museum, hoping to preserve the dwindling museum for all of Denver's citizens. The process of becoming a museum was successful, officially opening on 1 December 1994. This date is significant in that it coincided with the 100th birthday of Francis Lowry, for whom Lowry Field had been named.

The WORA&SM now had a running start, with twenty aircraft already conveniently located on site, many with historic significance. Like all newly formed groups operating on a shoestring, things went downhill for a time. Then, with a staff of very dedicated volunteers vowing to keep the museum going and operating in the black, things improved. Today the WORA&SM has a very enthusiastic paid staff and many plans for expansion.

The following aircraft are on display in the WORA&SM:

Alexander Eaglerock (Serial No.), Reg.No. NC2568. 1926 biplane built in Colorado
Beechcraft UC-45 Expeditor (Serial No. 42-37496)
Boeing B-52B Stratofortress (Serial No. 52-005)
Cessna U-3A Administrator "Blue Canoe" (Serial No. 57-5894)
Chanute Glider
Christian Eagle homebuilt

Convair F-102A Delta Dagger (Serial No. 56-0984)

Douglas B-18A Bolo (Serial No.39-0025)

FP-404 Biplane

General Dynamics FB-111A Aardvark (Serial No. 68-00297)

Goodyear FG-1D Corsair (Serial No. 92050), 3 ¼' clipped off each wing, modified and equipped with a Buick engine for racing.

Grumman F-14A Tomcat (Serial No. 159829), USS John C. Stennis

Hang Glider

Kitfox Model I

KR-1 Sport Plane

Langley Aerodrome No. 4

Lockheed T-33A Shooting Star (Serial No. 56-1710)

Lockheed F-104C Starfighter (Serial No. 56-910)

Luenger Beta I

Martin EB-57E Canberra (Serial No. 55-4293)

McDonnell F-101B Voodoo (Serial No. 58-271)

McDonnell-Douglas F-4E Phantom II (Serial No. 66-0286), "Julie"

MiniMAX 110R

Moni Experimental Aircraft

North American F-86H Sabre (Serial No. 53-1308)

North American F-100D Super Sabre (Serial No. 56-3417)

Nord 3202 (Serial No. 62), Reg. No. N2254R. L'Armée d'Air (French Air Force), primary trainer.

Piasecki (Vertol) H-21B Shawnee Helicopter (Serial No. 55-4218) (painted as 34379)

Primary Glider

Republic RF-84K Thunderflash (Serial No. 52-7266)

Republic F-105D Thunderchief (Serial No. 60-508)

Rockwell B-1A Lancer (Serial No. 74-00160)

Schweizer SGS I-24 Glider

Vought A-7D Corsair II (Serial No. 69-6188)

Wizard, powered ultralite

Woody Pusher

Other exhibits:

Luke Skywalker's X-Wing fighter. Model from the movie "Star Wars."
Space Station module simulator.
Numerous vintage cars, gliders, home-built aircraft and aviation memorabilia.

Denver, CO. **Worldwide Aircraft**. Marty Batuva,
E-mail: **marty@worldwideaircraft**. PO Box 790, Bellevue, NE
68005. Tel: 1-800-884-9490; or 402-291-6559. Fax: 402-291-1318.
Internet: **www.worldwideaircraft.com**.

Worldwide Aircraft is a full-service aircraft recovery company,
which will disassemble, package, transport, and reassemble aircraft
worldwide. The company has experience with all types of aircraft. It
has relocated over 2,500 aircraft from priceless vintage (Douglas World
Cruiser) to modern jets (F-15's and F-18's). It also has extensive experience moving aircraft for warbird owners, museum collections, and
the military.

Worldwide Aircraft is the <u>ONLY</u> aircraft recovery company that
performs all phases of the relocation in-house. The individuals who
disassemble the aircraft are the same people who transport and reassemble them. Worldwide employees are responsible for the work from
start to finish.

Worldwide Aircraft is totally self-sufficient in the field. They bring
the appropriate tools, equipment, and support that are required, and
will achieve the goal of relocating aircraft at a price that is affordable.

Worldwide Aircraft takes great pride in the projects they have completed and the relationships they have built with their customers. If
they can be of any help in your planning or your projects, please call or
e-mail them. Ben Nattrass, Owner.

Durango, CO.

Consolidated Vultee BT-13A Valiant (Serial No. 2960, Reg. No. N4566V. Brian M. Cutter, 160 Brice Place, Durango, CO 81303-8135.

Elbert, CO.

Fairchild 24R46 (Serial No.), Reg. No. NC81281. Jamie S. Treat, Aircraft Restoration and Repair, 24201 David C. Johnson Loop, Kelly Airpark, Elbert, CO 80106. Tel: 719-536-0459. E-Mail: **JSTreat360@aol.com**.
North American AT-6G Texan (Serial No. 49-2897), Reg. No. N7197C. Edward Huber, 24961 Kolstad Loop, Elbert, CO 80106-9512.

Elizabeth, CO.

Goodyear FG-1D Corsair (Serial No. 92050), Reg. No. N194G, James R. Axtell, 1066 Osage Street, Elizabeth, CO 80107-8436. Moved, presently unlocated.

Englewood, CO. **Centennial Airport**, 7800 South Peoria St., Englewood, CO 80112.

Aero L-39ZA Albatros (Serial No. 332634), Reg. No. N390ZA. 330 S Zuni St., Englewood, CO 80110-2087.
Cessna O-2A Skymaster (Serial No. 69-7669), Reg. No. N51680, all black, USAF colors. Patrol Squadron Colorado, 6838 Alpine Drive Lane, Parker, CO 80134.
Douglas DC-3 (Serial No.), WORA&SM, maintained in flying condition
Hunting (Percival) P.84 Jet Provost Mk. 5A (Serial No. XN471), Reg. No. N471XN. Craig Wingert, 1412 Eagle View Drive, Colorado Springs, CO 80909-2918.

Lockheed/Canadair T-33A Shooting Star (Serial No. 160), Reg. No. N221SF. Flight Test Research Inc., 6200 Plateau Drive, Englewood, CO 80111.

North American T-28B (Serial No. 137736), Reg. No. N28TP. John L. Breit, 12260 Control Tower Road, Englewood CO 80111.

Piper J3 Cub (Serial No. 18520), Reg. No. N98349, Steven J. Beckman, 5753 S. Orleans St., Aurora, CO 80015-3501.

Piper L-4J Cub (Serial No. 44-13640), Reg. No. N5985V, "Patsy"

North American SNJ-3 Texan (Serial No. 786987), Reg. No. N7648E. Robert A. Estock, 2752 S. Irving St., Denver, CO 80236-2510.

North American AT-6G Texan (Serial No. SA060), Reg. No. N48JC, Centennial, 12720 East Control Tower Road, Englewood, CO 80112.

Erie, CO.

Grumman G-44A Widgeon (Serial No. 1429), Reg. No. N945JD. Wayne C. Mulher, 185 Piper Drive, Erie, CO 80516.

Evergreen, CO.

North American AT-6A Texan (Serial No. AJ0968), Reg. No. N4802E, Rocky Mountain Group Technologies Inc., 3253 Elk View Drive, Evergreen, CO 80439-7973.

Falcon, CO. Meadowlake Airport, Cross Taxi Way E, Hangar #3, Falcon, CO. Aircraft Restoration. Tel: 719-70-0782.

Flagler, CO.

Lockheed T-33A Shooting Star (Serial No. 57-0587), property of the USAF Museum, location unconfirmed.

Martin YCGM-13A (TM-76A) Mace missile (Serial No. 58-1463)

Fort Carson, CO.

Beech C-45F Expeditor (Serial No. 44-87018), Reg. No. N6645C, Sylvan V. Vick, CO 45th Aviation Bn, Fort Carson, CO 80913.

Fort Collins, CO.

Hawker Sea Fury (Serial No.) Owned by the Whittington brothers, World Jet Center, Florida.
North American AT-6 Texan (Serial No.)
North American F-51H Mustang (Serial No.)

Fort Collins, CO. Evergreen Ventures.

Evergreen Ventures is restoring a Messerschmitt Bf 109 G-6 (Serial No. 610824), which was purchased by Doug Arnold's Warbirds of GB. It was recovered from the Yugoslav Aeronautical Museum. A Messerschmitt Bf 109 G-14 (Serial No. 610937) is also being restored after being purchased from Doug Arnold's Warbirds of GB. It was captured near Munich in Germany in 1945. For more information visit the Classic Warbirds website.

Fort Collins, CO. Q G Aviation Of America, Aircraft Restoration, 140 Racquette Dr
Ft Collins, CO. Tel: 970-221-5461.

Fort Collins, CO. Vintage Aircraft Ltd Inc., 2050 Airway Ave, Ft Collins, CO 80524-2708. Tel: 970-493-3477.

Fort Lupton, CO.

North American T-28C Trojan (Serial No. 140519), Reg. No. N9019N. Robert A. Resling, 7507 Weld County Road 39, Fort Lupton, CO 80621.

Frederick Firestone Airfield, CO.

Fairchild PT-19 Cornell (Serial No.)
Fokker D.VII (Serial No.), fuselage, replica
Sikorsky CH-34 Choctaw (Serial No.)
Taylorcraft L-2 Grasshopper (Serial No.)
Collection of various aviation parts and material, including a B-29 tug accumulated by Jim McDonald.

Franktown,CO.

De Havilland DHC-2 Beaver (Serial No. 981), Reg. No. N411MC, Reed Hollow Ranch LPA, 441 Russellville Road, Franktown, CO 80116-8901.

Golden, CO.

Bell UH-1H Iroquois (Serial No. 65-10061), Reg. No. N22SD. Jefferson County Sheriff's Department, 200 Jefferson County Parkway, Golden, CO 80401.
Piper L-21B Super Cub (Serial No. 18-3926), Reg. No. N33MT. Nick Saum, Box 956, Golden, CO 80402.

Grand Junction, CO.

Bell UH-1H Iroquois (Serial No. 67-17534), Reg. No. N3276M. Woodstone Corp., 3320 North Ridge Drive, Grand Junction, CO 81506-1926.
Grumman F-11 Tiger (Serial No. 141796). Blue Angels colors, mounted on a pylon at Walker Field Airport.
Grumman A-6E Intruder (Serial No. 154131). USN colors, mounted on a pylon at Walker Field Airport.

Grand Junction, CO. **Commemorative Air Force**, Rocky Mountain Wing, Grand Junction, CO. Tel: 303-244-9100. Contact: Bob Thompson, Tel: 970-921-4088.
Email:**rvtglt@gj.net**.

Douglas B-26C Invader (Serial No. 28880), Reg. No. N202R, "Miss Murphy", Rodney G. Huskey, Lonestar Enterprises, PO Box 55331, Grand Junction, CO 81505-5025.
Grumman TBM-3E Avenger (Serial No. 53503), Reg. No. N53503.
Piper J-3 Cub (Serial No.), Eric Baldwin.

Greely, CO. **Weld County Airport**

North American T-38A Talon (Serial No. 60-0586). Re-painted white in November 2000 and under restoration at the airport.
Cessna O-2 Skymaster (Serial No.), 7 aircraft, J.W. Duff Aircraft Salvage, Denver.

Greely, CO. **Colorado Air National Guard Base, Greely**, CO

Vought A-7 Corsair II, (Serial No.)

Homelake, CO. **American Legion Post 51**

Northrop Radioplane XQ-4 Missile (Serial No. 55-3233)

Johnstown, CO.

Hiller OH-12G Raven (Serial No. 1527), Reg.No. N603RA. Ag Air Inc., PO Box 606, Johnstown, CO 80534-0606.

Lakewood, CO.

Ryan ST-A/PT-22 Recruit (Serial No. 115), Reg. No. N14983. Gordon F. Autry, 1820 Winfield Drive, Lakewood, CO 80215.

Larkspur, CO.

North American T-28C Trojan (Serial No. 140464), Reg. No. N464SB. Perry Park Investments, Ltd., PO Box 98, Larkspur, CO 80118-0098.

La Salle, CO.

Bell UH-1F Iroquois (Serial No. 66-1216), Reg. No. N22064. Associated Aerial Applicators, 23482 WCR 48, La Salle, CO 80645.
North American AT-6C Texan (Serial No. 88-12018), Reg. No. N60380, Low Level Dusting Company Inc., 119 2nd Street, La Salle, CO 80645.

Littleton, CO.

North American AT-6G Texan (Serial No. 49-3038A), Reg. No. N66TY. Glenn R. Jones, 6585 S. Cherry Way, Littleton, CO 80121.
Piper J3 Cub (Serial No. 19299), Reg. No. N6134H. Mark E. Kirkegaard, 6741 South Marion Circle W, Littleton, CO 80120.

Longmount, CO.

Beech C-45G Expeditor (Serial No. AF-220), Reg. No. N9660C, Sports Travel Inc., 7214 Cardinal Lane, Longmont, CO 80501.
North American AT-6G Texan (Serial No. SA079), Reg. No. N26WR, Warren K. Rempel, 2745 Grinnel Drive, Longmount, CO 80503.
North American T-28A Trojan (Serial No. 49-1580), Reg. No. N281TT. Ken A. Buckspan, 3920 Iron Crescent, Longmount, CO 80503-8306.

Mead, CO.

Yakovlev Yak-11 (Serial No. 458519), Reg. No. N11YH. Jim H. McKinstry, 3750 Weld County Road, Mead, CO 80542.

Monument, CO.

Lockheed T-33A (Serial No. 49-0884), Reg. No. N652, Jim Cullen, Box 771, Monument, CO 80132. May have been sold, presently unlocated.
North American SNJ-4 Texan (Serial No. 88-13763), Reg. No. N6425D. Jerrold E. Wannemacher, PO Box 1154, Monument, CO 80132.

Montrose, CO.

Sikorsky UH-19D Chickasaw (Serial No. 57-5958), Reg. No. N555DT. Delbert Topliss, RR 2, PO Box 124, Montrose, CO 81401.
Vought A-7D Corsair I1 (Serial No. 70-1055). This aircraft is mounted on a pylon at Montrose Airport.

New Liberty, CO.

Bell UH-1H Iroquois (Serial No. 69-15911), Reg. No. N3276T. Woodstone Corporation, 1856 8 Road, New Liberty, CO 81525.

Ouray, CO.

Bell 47G-2 Sioux (Serial No. 1973), Reg. No. N2871B. Jack Summers, PO Box 43, Ouray, CO 81427-0043.

Pagosa Springs, CO.

Bell 47G Sioux (Serial No. 5503), Reg. No. N4475. Kenneth J. Tooker, 2770 Ranchland Drive, Pagosa Springs, CO 81147-9824.
Bell 47G-3B Sioux (Serial No. 3267), Reg. No. N1509L. Rotors West Inc., 2770 Ranchland Drive, Pagosa Springs, CO 81147-9824.
Grumman G-44 Widgeon (Serial No. 1372), Reg. No. N92L. John M. Huft, 2400 Preservation Place, Pagosa Springs, CO 81147-8104.
North American P-51D Mustang (Serial No. 44-73518N), Reg. No.

N5483V. Whittington Brothers Inc., 1881 St., Rd 84, Fort Lauderdale, FL 33315

Parker, CO.

Piper J3 Cub (Serial No. 19881), Reg. No. N6676H. Rob Shriner Jr., 3234 East Cottonwood Ave., Parker, CO 80134-9623.

Platte Valley Airpark, CO.

Lafayette Escadrille (WWI flying replica aircraft, privately owned). The Lafayette Foundation is a collection of WWI warbirds, vintage and classic airplanes. Info: Linda Fore 719-599-3869. Email: **cessna185f@aol.com**.
At Platte Valley Airpark, northeast of Denver, the foundation hangars replica WWI aircraft, including a Sopwith Pup and Fokker Dr.1 (Dreidecker), a tri-plane. The Fokker Dr.1 is similar to the one flown by Baron von Richthofen, "the Red Baron." The collection also includes "Aquarius," a Fokker D.VII built by Dr. Parks, and two S.E.5s.
Of six WWI replica aircraft, four are currently in operational flying status. An S.E.5 will be built, one half in fabric, and the other half in see-through material, so the curious can see how these planes are constructed.

Fokker D.VII (Serial No.), full scale replica
Fokker Dr.I (Serial No.), full scale replica
Royal Aircraft Factory SE-5A (Serial No.), 5/8 scale replica
Royal Aircraft Factory SE-5A (Serial No.), 5/8 scale replica
Sopwith Pup (Serial No.), 5/8 scale replica

Pueblo, CO. **Pueblo Weisbrod Aircraft Museum**, Pueblo Memorial Airport, 31001 Magnuson Avenue, Pueblo, CO, 81001. Tel: 719-948-9219; FAX: 719-948-2437. Internet: **http://www.pwam. org/museums.html**.

The PWAM unique to the state of Colorado, and is a tribute to American military airmanship and is the largest collection of vintage aircraft on the Eastern Range of Colorado.

On 30 June 1972 the first aircraft, a Douglas A-26 Invader arrived on site, followed soon thereafter by a Lockheed F-80 Shooting Star. The other aircraft currently on display are on loan from the various military services with the exception of the McDonnell Douglas F-101, recorded as having a civilian owner.

Located six miles east of the city on Highway 50 at the Pueblo Memorial Airport, the aircraft display and the International B-24 Memorial museum occupy space on what was the Pueblo Army Air Base during WWII. The museum is open to the public seven days a week. Admission is $4.00 per person.

Members of the Pueblo Historical Aircraft Society, a volunteer group of ex-military and civilian personnel, manage and operate the aircraft display as well as performing the repair and restoration of the aircraft.

The Society is open to all who wish to join and become a part of the team dedicated to preserving the aviation history of the city of Pueblo. The organization may be contacted by writing to:

Pueblo Historical Aircraft Society, 31001 Magnuson Ave., Pueblo, Co. 81001. Tel: 719.948.9219. E-mail: **service@pwam.org**.

The Pueblo Weisbrod Aircraft Museum, home of the International B-24 Museum is building a database of former B-24/PB4Y-1 Liberator and PB4Y-2 Privateer aircraft crew members. The database will help veterans locate friends and serve as a source for historical research. Veterans or family members of veterans wishing to add the names of former crew members to the data base may contact the museum and request a B-24 Crew Member Form. The aircraft are on display in the museum:

Bell 47 UH-13J Sioux Helicopter (Serial No. 14941), Reg. No. N3682
Bell UH-1 Iroquois (Huey) Helicopter (Serial No. 72-21508)

Boeing/Stearman PT-13D Kaydet (Serial No.), Reg. No. N1944S. Owned by Rudy Eskas since 1979. 9-cylinder version.

Boeing B-29A Superfortress (Serial No. 44-62022)

Boeing B-47E Stratojet (Serial No. 53-02104)

Chance Vought V-346 F7U-3 Cutlass (Serial No. 128451), project

Convair HC-131A Samaritan (Serial No. 5794)

Douglas C-47/R4D Skytrain (Serial No. 42-93328)

Douglas A-26C Invader (Serial No. 44-35892)

Douglas F-6A/F-4D Skyray (Serial No. 134936)

Fairchild C-119 Flying Boxcar (Serial No. 131688)

Grumman F9F-8 Cougar (Serial No. 138876)

Grumman F11F-1 Tiger (Serial No. 141853)

Lockheed P2V-5 Neptune (Serial No. 128402), painted as 151353, VP-19

Lockheed F-80A Shooting Star (Serial No. 137939), painted as 91872

Lockheed T-33B (Serial No. 137939), USMC colors

Lockheed RB-37 Ventura (Serial No. 44-449), AJ311

McDonnell Douglas A-4D-2 Skyhawk (Serial No. 147702), 512

McDonnell F-101A Voodoo (Serial No. 53-02418)

North American F-100D Super Sabre (Serial No. 56-3299), Thunderbird colors

North American T-28C Trojan (Serial No. 140064)

North American RA-5C Vigilante (Serial No. 151629), RVAH-3, 314

Piasecki CH-21C Workhorse Helicopter (Serial No. 0-34347)

Republic F-84C Thunderjet (Serial No.47-01562), project

Sikorsky SH-34J/S-58 Choctaw (Serial No. 148002), US Naval Air Systems Command

Vought F-8 Crusader (Serial No. 145349)

Mass Transit Research Museum Vehicles on display at the museum:

Garret Linear Induction Motor

Grumman Tracked Levitation

Rohr Tracked Air Cushion vehicle

(The American Association of Railroads has a High-Speed Test Track nearby)

Support Equipment on display at the museum:

1942 Ford truck aircraft refueler
Amber Three Beacon Airport Tower

Rangely, CO.

Hiller OH-23C Raven (Serial No. 64-15144), Reg. No. N5578Z. Lloyd B. Roper, PO Box 1264, Rangely, CO 81648.

Salida, CO. B&J Flying Service, Aircraft Restoration, 9275 County Rd #140, Salida, CO 81201-9437. Tel: 719-539-7955.

Salida, CO. Jensen Aircraft, Aircraft Restoration, 9225 County Rd #140 Salida, CO 81201. Tel: 719-539-7050.

Schriever AFB, CO. Internet: **www.schriever.af.mil**. Schriever AFB is located 10 miles east of Colorado Springs and houses the 50th Space Wing, Space Warfare Center and Joint National Test Facility.

South Fork, CO.

Beech C-45H Expeditor (Serial No.), 10 aircraft.
North American AT-6 Texan (Serial No. 42-44722), Reg. No. N3660V, SLV Crop Care Inc., 457 Rio Grande Drive, South Fork, CO 81154.

Steamboat Springs, CO

Hiller UH-12E Raven (Serial No. 5029), Reg. No. N529HA. Edward W. Neish, 29765 Routt County Road 18, Steamboat Springs, CO 80487.

Stirling, CO.

North American T-28C Trojan (Serial No. 140578), Reg. No. N8039S. Noah C. Kimball, 832 Hawthorne Crescent, Stirling, CO 80751.

Watkins, CO.

Piper J3 Cub (Serial No. 5734), Reg. No. N32950. Richard L. Lepka, 28580 East Colfax Ave., Watkins, CO 80137.
Vought A-7D Corsair II (Serial No. 71-0337), Airport Authority Static Display.

Wray, CO. Triangle Aviation, Aircraft Restoration, 36391 US Highway 385, Wray, CO 80758. Tel: 970-332-4656.

Alphabetical list of Warbird Survivors in Colorado

Aero L-39 Albatross

The L-39 is a tandem two-seat basic and advanced jet trainer powered by one 3,792-lb thrust Walter Titan turbofan (Ivchenko AI-25-TL built under license by Motorlet). It has a maximum speed of 485-mph at 19,685' and a service ceiling of 37,730', a range of 528 miles and can be armed with up to 2,425-lbs of weapons on four underwing hardpoints. The Albatros was designed before the Soviet armed intervention in Czechoslovakia in 1968, and is a follow-on trainer to the L-29 Delfin. *(The Complete Encyclopedia of World Aircraft)*.

Alexander Model A-14 Eaglerock

The Eaglerock biplane, made famous by barnstormers during the 1920s, was manufactured in what is now downtown Englewood, Colorado, and later in Colorado Springs, by the Alexander Aircraft Company. Barnstormers landed the Eaglerock in farm fields across rural America in the 1920s and '30s, giving rides in these "new flying machines" to the brave souls willing to take the risk of flight. Ten-minute rides sold for 50 cents to a dollar.

The Eaglerock's builders, the Alexander brothers, began establishing themselves as businessmen by selling street advertising. After a brief detour into the chicken raising business, they founded the Alexander Film Company, specializing in screen advertising. The success of the Alexander Film Company rapidly necessitated an increase in staff, and a relocation from their original home in the State of Washington to

Englewood, Colorado. The Colorado location was more central to their film advertising business.

It was J. Don Alexander who came up with the idea of equipping his growing sales force with airplanes. This would serve two purposes: first, it would attract attention, and second, it would expedite distribution of the advertising films. The first plane, purchased by older brother S. Don Alexander, was a 1920 Laird "Swallow," powered by an OX5. When the Swallow arrived in Denver it landed at Lowry Field, located at 38th and Daliah Streets in Denver. The next additions to the Alexander aircraft fleet were Longren biplanes.

J. Don Alexander wanted to purchase some forty to fifty planes for his salesmen. However, no one, not even the government, was buying that many aircraft in the 1920s, so the existing aircraft manufacturers would not take Mr. Alexander's proposal seriously. This prompted him start his own aircraft manufacturing company.

Colorado's Alexander Aircraft Company built over 900 airplanes in the 1920's and 30's. The factory was initially located at 3385 South Broadway, just south of the Denver, Colorado City limits.

One of the early Alexander Aircraft Company employees, a 19-year-old youngster name Al, had a job at the factory building aircraft wing ribs. Al was instrumental in the early success of the Eaglerock, his suggestions resulted in some significant design changes to one of the Eaglerock airplanes under construction. This youngster, Al Mooney, went on to found his own successful aircraft company.

The Eaglerock was popular with well-known aviators of the day, including a noted early flyer known to some as "Slim." The Eaglerock was in the running with Bellanca as the aircraft-of-choice for Charles A. Lindbergh's historic New York to Paris flight. It was only due to both factories already being buried with production orders that "Lucky Lindy" ended up with the successful Ryan monoplane.

With their aircraft business expanding rapidly, and new designs being considered, expansion of the Eaglerock factory was necessary. Forced out of the Denver area by a landowner's refusal to sell the land needed for expansion, Alexander Aircraft relocated to Colorado Springs, Colorado.

The Alexander Aircraft Company went on to build the unsuccessful "Alexander Transport," a high wing, seven-passenger monoplane. However, other more successful models followed. In the 1928-1929 time frame the Alexander Aircraft Company was the largest aircraft manufacturer in the world, with the capacity of manufacturing eight airplanes a day.

It is estimated there are some 24 Alexander aircraft still around as fliers, on display, or in the process of being restored. At least one Alexander Bullet is being constructed in Longmont, Colorado. The oldest known surviving Alexander Eaglerock, a 1926 Long Wing, Reg. No. NC2568, is on display at the Wings Over the Rockies Air and Space Museum. Another Alexander Aircraft Eaglerock biplane is on display at Denver International Airport, at the West End of United Airlines' Concourse B. This is the aircraft that appears on the Museum's logo. *(Information courtesy of Ronald E. Newberg, WORA&SM).*

Members of the Antique Airplane Association of Colorado and Bayport, New York restored Eaglerock Reg. No. NC205Y over 25 years, ending in 1989. This aircraft is now on display overhead at DIA, and was a gift to the City and County of Denver to honor the pioneering spirit of the West during the Golden Era of Aviation. NC205Y is painted in red and gray livery, and is located between Gates B-22 and B-24.

Avro CF-100 Mk. 5C Canuck
(HAS Photo)

The Avro CF-100 twin-jet interceptor was equipped with two 7,275-lb Avro Orenda 11 or Orenda 14 turbojet engines. To improve the Canuck's high-altitude performance, its wing-span was increased by 6' and a larger tailplane was installed. It was only armed with the wing-tip missile pods. The Mk. 5 had a maximum speed of 1,046 km/h, a service ceiling of 54,000', and a range of 3,220 km. They entered RCAF service on 18 June 1955.

The Canuck was one of Canada's great aviation success stories. It is the only fighter ever designed and built in Canada, and to go into mass-production. It served both the RCAF and the Canadian Forces (CF) for over 30 years. It began its service as a front-line fighter in Canada and overseas in NATO posts in Europe, and ended its career as a trainer. Often called the "Clunk," the CF-100 was a highly reliable and competitive fighter. 53 were supplied to the Belgian Air Force. 50 Mk. 4Bs were also later converted to Mk.5 standard. *(Information courtesy of the CF Archives)*.

The Canuck shown above is on display in the Peterson AFB Air Park, beside the Peterson Air and Space Air Museum). This CF-100 is an electronic counter-measures version. It was flown to Peterson AFB on 27 May 1976 by Lt Col E.G. Francis and Major R.A. Willhauk with No. 414 Squadron, North Bay, Ontario. The formal transfer and dedication of the CF-100 was made in July 1976. *(Information courtesy of the PA&SM).*

Beechcraft C-45F/G/H Expeditor

The Beech Model 18 is a twin-engine light transport powered by a pair of 450-hp Pratt & Whitney R-985-AN-14B radial piston engines. The aircraft had a maximum speed of 220 mph, a service ceiling of 21,400', and a range of 1,530 miles. *(The Complete Encyclopedia of World Aircraft).*

One of the most famous twin-engine aircraft ever built, the Beech was used during WWII for communications work and instrument flying training. After the war, the Expeditor was used as a basic multi-engine trainer. First flown in 1937, the aircraft was used to train pilots and radio officers, transport VIPs, and as a general transport aircraft. Later in its career, the aircraft was often used in search and rescue (SAR) missions. The C-45 was the WW II military version of the popular Beechcraft Model 18 commercial light transport.

During 1951-52, USAF UC-45E, T-7 and T-11s in service with the USAF were remanufactured to zero-time condition and modernized. These aircraft were designated C-45G and C-45H. At the same time the US Navy SNB-5 and SNB-5P aircraft were modified, and in 1962 redesignated TC-45J for training and RC-45J for photographic missions.

Beech built a total of 4,526 of these aircraft for the Army Air Forces between 1939 and 1945 in four versions, the AT-7 Navigator navigation trainer, the AT-11 Kansan bombing-gunnery trainer, the C-45 Expeditor utility transport and the F-2 for aerial photography and mapping. The AT-7 and AT-11 versions were well-known to WW II navigators and bombardiers, for most of these men received their training in these aircraft. Thousands of AAF pilot cadets also were given advanced training in twin-engine Beech airplanes. *(Information courtesy of the USAF Museum).*

An Expeditor is on display inside the WORA&SM.

Bell Model 47/UH-13J Sioux Helicopter

The Bell Model 47 helicopter is a three-seat light utility helicopter powered by an Avco Lycoming VO-435-B1A flat-six piston engine. It had a maximum speed of 105 mph and a range of 250 miles. *(The Complete Encyclopedia of World Aircraft).*

The Model 47 has been around longer than any other helicopter, with the first aircraft having been flown on 8 December 1945. In March 1946 it was certified by the United States as a commercial helicopter and its production run of over twenty years (in both military and civilian circles) was the longest in history. Italy and Britain built a number of variants. Unique to the Bell design was the distinctive gold-fish-bowl cockpit that offered a wide-range of pilot visibility. The counter-balanced see-saw rotor system of the Sioux created the distinctive clop-clop noise associated with helicopters and from which the nickname choppers for helicopters originated.

Procured by the Navy in 1947, the HTL-4s were used as trainers and as the vehicle for gaining experience in rotary-wing operations. Until retired in April 1964, they were used in the Apollo program to familiarize astronauts with its auto-rotation sink rate, which was similar to the descent rate of the Apollo Lunar Excursion Module. Some HTLs were fitted with pontoons for use in shipboard operations while others were equipped with external stretcher pads on either side of the cockpit and used by the Marines in Korea as medical evacuation helicopters. It was the latter variant of the Sioux that became famous with its portrayal in the M*A*S*H television series. *(Information courtesy of the National Museum of Naval Aviation).*

A Bell 47 is on display in the PWAM.

Bell UH-1H Iroquois Helicopter

The Bell UH-1H is a general-purpose utility helicopter powered by one 1,400-hp Avco Lycoming T5313B turboshaft. This helicopter had a maximum speed of 127 mph, a service ceiling of 12,600' and a range of 318 miles. *(The Complete Encyclopedia of World Aircraft).*

The Iroquois evolved from a 1955 Army competition for a new utility helicopter. The Army employed it in various roles, later including that of an armed escort or attack gunship in Vietnam. The USAF, USN, and USMC eventually adopted the model, as did Canada, Brazil, and West Germany. The initial Army designation was HU-1, which led to the common unofficial nickname of Huey. It was redesignated in 1962 as the UH-1 under a tri-service agreement.

USAF orders for the Huey began in 1963 for UH-1Fs, intended for support duties at missile sites, and for TH-1Fs for instrument and hoist training and medical evacuation. The HH-1H incorporated a longer fuselage and larger cabin for a crew of two and up to eleven passengers or six litters. The USAF ordered these in 1970 as local base rescue helicopters to replace the HH-43 Huskie. The first of the USAF's UH-1Ns, a twin-engine utility version capable of cruising on one engine, was obtained in 1970. *(Information courtesy of the USAF Museum).*

An Iroquois is on display at the PWAM; another can be found at Butts Airfield, Fort Carson, and several others are at Buckley AFB.

Bell Model 209 AH-1F & G Huey Cobra Helicopter
(HAS Photo)

The Cobra is an attack/close support helicopter powered by a Pratt & Whitney Aircraft of Canada T400-CP-400 twin-engined turboshaft unit. It has a maximum speed of 207 mph, a service ceiling of 12,450', and a range of 359 miles. First flown in September 1965, the Cobra is armed with one a six-barreled Minigun, or one M-197 three-barrel 20mm cannon carried in a tactical armament turret faired into the undersurface of the front fuselage. It can also carry up to 2,200-lb of weapons including the XM-157 seven-tube or XM-159 19-tube 2.75in rocket pods on four underwing racks. The Huey Cobra presents a very slim profile, particularly in its most-deadly position, facing an enemy head-on. *(The Complete Encyclopedia of World Aircraft)*.

The AH-1F Cobra shown above is on display outside the Head-quarters building for the 4th Squadron, 3rd Armored Cavalry Regiment, Building 2350, Titus and Barkely Street, Fort Carson. An AH-1F is also on display outside the Main Gate to Fort Carson, next to a statue of Kit Carson.

Bell OH-58A Kiowa

The Kiowa helicopter is a small, single-engine, single-rotor (two-bladed) general-purpose variant of the Kiowa helicopter acquired to support the Army in the observation, reconnaissance, command and liaison, target acquisition, and adjustment of fire roles. The civilian variant of this helicopter is the Bell Jet Ranger. The Kiowa is powered by one 420-hp Allison 250-C20B turboshaft, with a maximum speed of 134 mph, a service ceiling of 13,500' and a range of 378 miles. *(The Complete Encyclopedia of World Aircraft).*

Following certification by the Federal Aviation Administration in 1966, the sleek, streamlined design of the Bell Jet Ranger made it an instant hit in the civilian market with about 300 being built by mid-1968. It was at this time that the Army and Navy took a renewed interest in the Bell product. The Army named Bell as the winner of a reopened design competition for a light observation helicopter and ordered 2,200 production versions (designated OH-58). Deployed to Viet Nam as early as 1969, the OH-58 served as a scout helicopter. There were 2,200 Kiowas produced. *(Information courtesy of the National Museum of Naval Aviation).*

A Kiowa is on display inside the Main Entrance to Butts Airfield, Fort Carson.

Boeing/Stearman Model 75/PT-13D/N2S-5 Kaydet
(HAS Photo)

The Kaydet is a two-seat primary biplane trainer used during WWII. The Model PT-13D was powered by one Avco Lycoming R-680-17 radial piston engine. The aircraft had a maximum speed of 124 mph, a service ceiling of 11,200', and a range of 505 miles. *(The Complete Encyclopedia of World Aircraft)*.

The need for a relatively inexpensive and easy to fly trainer for the Army and Navy was met by the Boeing Aircraft Company which produced the Stearman Model 75 (Army designation PT-13, PT-17, PT-18, PT-27, and Navy designation N2S), in 1935. The N2S-5 Kaydet was the US Navy equivalent of the PT-13D Model 75 Kaydet. The Stearmans were relatively similar, except a Lycoming engine powered the PT-13 version, a Continental engine powered the PT-17, and the PT-18 was powered by a Jacob's engine.

The PT-27 was outfitted for use in Canada and was similar to the PT-17 but had an enclosed cockpit and heating, night flying equipment, blind-flying hood and instruments. Production of the Kaydet ended in February 1945 with 10,346 being built. Many are still in use today, including the one owned by Rudy Eskra (shown above) which he flies out of the PWAM. *(Information courtesy of the PWAM Museum).*

Boeing B-29A Superfortress
(HAS Photo)

The Superfortress was a long-range strategic bomber and reconnaissance aircraft. The B-29 was introduced in June 1944, and at the time represented a tremendous leap forward in technology. With four powerful 3,500-hp Pratt & Whitney R-4360-35 Wasp Major or four Wright R-3350-23-23A/-41 Cyclone 18 turbo-charged radial piston engines, it had the speed of many fighters (358 mph). The aircraft had a service ceiling of 31,850' and a range of 3,250 miles. The B-29 was heavily armed with two .50 cal machine-guns in each of four remotely-controlled power-operated gun turrets and three .50 cal machine-guns and one 20mm cannon in the tail turret. It could carry a bomb-load of 20,000-lbs. *(The Complete Encyclopedia of World Aircraft).*

The B-29 was the largest and most sophisticated bomber to enter combat in WWII. First flown in 1942 with initial production models revealed in 1943. Although plagued by engine problems early on it was an extremely strong, stable and efficient ac. A favorite with crews, due to pressurized and heated crew compartments. It had sophisticated RADAR and defensive armament. It served well in the Pacific during WWII, in Korea and then in the Strategic Air Command. Several versions were produced including an aerial re-fueling tanker. It was used as a launch platform for experimental super sonic aircraft. A total of 3979 aircraft were built.

The Superfortress primarily operated in the Pacific battle area, where it played a central role in the strategic campaign against Japan, which culminated in the dropping of Atomic bombs on Hiroshima (by the B-29 Enola Gay) and Nagasaki (by the B-29 Bocks Car). The Boeing B-29 was designed in 1940 as an eventual replacement for the B-17 and B-24. The first one built made its maiden flight on 21 September 1942. In December 1943 it was decided not to use the B-29 in the European Theater, thereby permitting the airplane to be sent to the Pacific area where its great range made it particularly suited for the long over-water flight required to attack the Japanese homeland from bases in China. During the last two months of 1944, B-29s began operating against Japan from the islands of Saipan, Guam and Tinian. With the advent of the conflict in Korea in June 1950, the B-29 was once again thrust into battle. For the next several years it was effectively used for attacking targets in North Korea. *(Information courtesy of the USAF Museum).*

The Pueblo Museum's B-29 "Peachy" (shown above), is named in honor of all the crews who fought in the Pacific Theater. A B-29 by that name was piloted by a native of Pueblo, Lt. Robert T. Haver, who named it with his pet name for a younger sister. The original "Peachy" flew 35 combat missions into enemy territory from Tinian island, central Pacific chain of the Mariannas. This aircraft was donated to the museum in 1976 by the Naval Weapons Center at China Lake CA.

(Information courtesy of the PWAM). A second B-29 was on display at Lowry AFB until it was flown to Seattle, Washington, where it is now on display outside the Seattle *Museum of Flight.*

Boeing Model 367 KC-97L Stratofreighter
(HAS Photo)

The Stratofreighter is a long-range military transport or inflight-refueling tanker powered by four 3,500-hp Pratt & Whitney R-4360-59B radial piston engines. It had a maximum speed of 375 mph, a service ceiling of 30,200', and a range of 4,300 miles. *(The Complete Encyclopedia of World Aircraft).*

The Stratofreighter had much in common with the B-29 Superfortress including the entire wing and engine layout. The lower structure was basically the B-29 and so was the tail unit, but the upper structure incorporated a bubble shaped fuselage. The transport version could carry a 53,000-lb payload. The KC-97G tanker was the most numerous variant with 592 being built. *(Information courtesy of the USAF Museum).*

The KC-97 was derived from the B-29 Superfortress bomber, using the same wing, engine, and tail. The new fuselage had a double-bubble configuration. The C-97 served a varied mission when active; in addition to both troop and cargo transport and tanker roles, it was used for medical evacuation, Search And Rescue (SAR) and by the Israeli Air Force, as an Electronic Countermeasures (ECM) platform. Boeing's KC-97 Stratofreighter aerial tanker was an instrumental factor in providing SAC with genuine intercontinental capability. Over 800 Stratofreighters were received between 1951 and 1956, by which time no less than 36 squadrons used the type. During their lifetime, the KC-97s transferred millions of gallons of fuel in air-to-air operations. Operations involving air refuelings average 2,880 contacts per week, or one aerial servicing every three and a half minutes. *(Information courtesy of the Strategic Air and Space Museum).*

The Boeing Model 367 KC-97L Stratocruiser (Serial No. 53-283), Texas Air National Guard, shown above, is on display at Solo's Restaurant, 1665N Newport Road (off Fountain, one block east of Powers), in Colorado Springs. Tel: 570-7656. Hours: Sun.-Thurs., 11 a.m.-9 p.m.; Fri.-Sat., 11 a.m.-10 p.m.

Boeing B-47E Stratojet
(HAS Photo)

The Stratojet is a strategic medium bomber powered by six 7,000-lb thrust (with water injection) General Electric J47-GE-25 or -25A turbojets. It has a maximum speed of 606 mph, a service ceiling of 40,500', and a range of 4,000 miles. It is armed with two 20-mm cannon mounted in a remotely-controlled tail turret, and can carry up to 20,000-lbs of bombs internally. *(The Complete Encyclopedia of World Aircraft)*.

The Boeing B-47, the world's first swept-wing bomber to be built in quantity, made its initial flight on 17 December 1947 and quantity deliveries began in 1951. When production ended in 1957, more than 1,200 Stratojets were serving with the Strategic Air Command at USAF bases throughout the world. By the late 1960s, the B-47 was obsolete and was removed from operational service. The B-47's thin flexible wing drooped when it was at rest. When it was carrying a bomb load in flight, the wings flexed in the opposite direction to provide a conventional dihedral wing. The B-47 normally carried a crew of three, pilot, copilot (who operated the tail turret by remote control), and an observer who also served as navigator, bombardier and radar operator. In the RB-47 reconnaissance version, the navigator also operated the camera equipment. *(Information courtesy of the USAF Museum)*.

The B-47 was the new look in high performance swept wing bombers, and became the backbone of the Strategic Air Command. The E model was the fifth version produced, and was first flown on the 30th of January 1953. The "Stratojet" was the first production model to use bicycle type landing gear. After being phased out of active service many were converted to QB-47 radio controlled drones for use in evaluation tests. The Pueblo Museum's B-47 E (shown above) was donated in 1979 by the US Naval Air Systems command and was flown into Pueblo from Davis Monthan AFB, AZ. *(Information courtesy of the PWAM)*.

Boeing B-52B Stratofortress

The B-52 is a six-seat long-range strategic bomber. Current H versions are powered by eight 17,000-lb thrust Pratt & Whitney TF33-P-3 turbofans. It has a maximum speed of 595 mph, a service ceiling of 55,000', and an unrefueled maximum range of 10,000 miles. It is armed with one remotely-controlled 20-mm Vulcan cannon mounted in the tail turret, and can carry up to 20 AGM-86 Air Launched Cruse Missiles (ALCMs), nuclear free-fall bombs, or a tremendous warlord of conventional bombs. *(The Complete Encyclopedia of World Aircraft)*.

The Stratofortress has been described as the "big brother" of its predecessor, the B-47, and it retains the same type of flexible wing. In 1958 the B-52 represented the cornerstone of the West's deterrence policy, with its ability to carry nuclear weapons to any target in the world. The H model is still in service. Before coming to Lowry AFB, the B-52B aircraft on display at the WORA&SM (shown above), served with the 93rd Bomb Wing (SAC) at Castle AFB, California. *(Information courtesy of the WORA&SM)*.

Boeing B-52D Stratofortress

Since it became operational in 1955, the B-52 has been the main long-range heavy bomber of the Strategic Air Command. It first flew on 15 April 1952. Nearly 750 B-52s were built when production ended in Oct. 1963, of which 170 were -Ds. The -Ds were modified to carry conventional bombs and Quail decoy missiles. The B-52 has set many records in its 25-plus years of service. On 18 January 1957, three B-52Bs completed the world's first non-stop round-the-world flight by jet aircraft, lasting 45 hours and 19 minutes with only three aerial refuellings en route. It was also a B-52 that made the first airborne hydrogen bomb drop over Bikini Atoll on 21 May 1956. In June 1965, B-52s entered combat when they began flying missions in Southeast Asia (SEA). By Aug. 1973, they had flown 126,615 combat sorties with 17 B-52s lost to enemy action. *(Information courtesy of the USAF Museum).*

The USAF Academy's B-52D "Diamond Lil," (shown above), flew over 200 combat missions including eight during "Linebacker II." Tail-gunner Airman 1ˢᵗ Class Albert C. Moore onboard Diamond Lil, was the only B-52 gunner to shoot down a MiG-21 Fishbed during the

Viet Nam War. It was flown to PAFB and moved to the Academy in 1984. *(Information courtesy of the USAF Academy).*

Boeing CIM-10A BOMARC Missile

The BOMARC ("BO" from Boeing and "MARC" from Michigan Aeronautical Research Center), originally designated as the XF-99 and IM-99, was a surface launched pilotless interceptor missile designed to destroy enemy aircraft. Propelled at launch by a rocket booster until it reached sufficient speed for its ramjets to operate, it was guided from the ground to the vicinity of its target at which time it came under control of an internal target seeker. Testing of prototypes began in 1952 and the -A series was declared operational in 1960. The improved -B series became operational in 1961 and had a range of 440 miles and a maximum altitude of 100,000 feet. It had more powerful ramjet engines and its solid-propellant booster permitted the almost instantaneous launch of a missile on alert. In 1969, Bomarc -Bs were operational at six USAF sites in the U.S. and two RCAF sites in Canada. Bomarc -As were phased out in the mid-1960s, but beginning in 1962 some were modified and flown as supersonic, high altitude target drones (CQM-10a). Complete phase-out of the Bomarc's air defense mission was completed in October 1972. *(Information courtesy of the USAF Museum).*

A Bomarc is on display at the PA&SM.

Cessna T-41A Mescalero

The T-41A Mescalero is a four-seat cabin monoplane and trainer powered by one 180-hp Continental O-360 flat-six cylinder piston engine. It is modeled on the Cessna Model 172, and 170 were ordered by the USAF in July 1964 and used by students at the USAF Academy with the designation T-41A. *(The Complete Encyclopedia of World Aircraft).*

A T-41A is on display at the USAF Academy airport close to Highway I-25 in Colorado Springs.

Cessna O-2 Skymaster
(HAS Photo)

The O-2 is a military version of the six-seat Cessna Model 337 Super Skymaster. Distinguished by twin tail booms and tandem-mounted engines, it features a tractor-pusher propeller arrangement. Derived from the Cessna Model 336, the Model 337 went into production for the civilian market in 1965. It is powered by two 210-hp Continental IO-360-GB flat-six piston engines, has a maximum speed of 206 mph, a service ceiling of 18,000', and a range of 1,422 miles. *(The Complete Encyclopedia of World Aircraft).*

In late 1966, the USAF selected a military variant, designated the O-2, to supplement the O-1 Bird Dog forward air controller (FAC) aircraft then operating in Southeast Asia. Having twin engines enabled the O-2 to absorb more ground fire and still return safely, endearing it to its crews. The O-2 first flew in January 1967 and production deliveries began in March. Production ended in June 1970 after 532 O-2s had been built for the USAF. The O-2A was extensively used in Vietnam.

Two series were produced: the O-2A and the O-2B. The O-2A is equipped with wing pylons to carry rockets, flares, and other light ordnance. In the FAC role the O-2A was used for identifying and marking enemy targets with smoke rockets, coordinating air strikes and report-ing target damage. The O-2B was a psychological warfare aircraft equipped with loudspeakers and leaflet dispensers. It carried no ord-nance. *(Information courtesy of the USAF Museum)*.

The O-2 shown above was parked at Centennial Airport in Denver. At least seven others which belong to Duff Aircraft Salvage are located at Greely Airport.

Cessna Model 310 U-3 Administrator "Blue Canoe"

The Cessna U-3 Administrator is a 5/6-seat cabin monoplane version of the Cessna 310 civilian aircraft powered by a pair of Continental IO-520-MB flat-six piston engines. The Model 310 has maximum speed of 238 mph, a service ceiling of 19,750', and a range of 1,765 miles.

Designated the L-27 until 1962, the U-3 was more commonly called the "Blue Canoe" or "Bluebird" for its blue and white color scheme. It was used for executive transportation, liaison duties, and to haul light cargo between military bases. *(Information courtesy of the WORA&SM)*.

One is on display in the WORA&SM.

Chance Vought V-346 F7U-3 Cutlass

The Cutlass is a single-seat carrier-based fighter powered by two 6,100-lb thrust Westinghouse J46-WE-8A after-burning turbojets. It had a maximum speed of 680 mph, a service ceiling of 40,000', and a range of 660 miles. The F7U did not have tail-planes. It was armed with four 20mm cannon and it was equipped with underwing attachments for rockets or other stores. The Cutlass was one of the aircraft used by the Blue Angels display team. *(Information courtesy of the National Museum of Naval Aviation)*.

Only three other Cutlass fighters are known to exist, (Serial No. Bu-129655), at the *National Museum of Naval Aviation* at Pensacola, Florida; (Serial No. Bu-129642), in the *Wings of Freedom Air & Space Museum* at Willowgrove NAS, Philadelphia, Pennsylvania, and an F7U Cutlass (Serial No. 129554), which is undergoing restoration at Paine Field, Everett, Washington; in addition to the remains of the project at the *Pueblo Weisbrod Air Museum.*

This rare bird has been reduced to a "hulk" and needs serious attention. The wings and fins have been detached, and the remaining fuselage is in sad shape, although the unusual mahogany wood lining around the intakes is still visible.

De Havilland D.H.112 Venom

The Venom is a single-engined jet-fighter bomber powered by one De Havilland Ghost 103 turbojet engine. It has a top speed of 587 mph, a service ceiling of 40,000', and a range of 705miles. It was armed with four 20-mm cannon, and could be equipped with two Firestreak air-to-air missiles, or two 1000-lb bombs or eight rockets.

The Venom was a follow-on development of the D.H.100 Vampire, using a thinner wing and powered by a more powerful turbojet engine. 150 were delivered to the RAF from May 1954 onward, and another 250 were supplied to the Swiss Air Force. *(The Complete Encyclopedia of World Aircraft).*

The Venom flying out of Broomfield is a former Swiss Air Force aircraft.

Convair F-102A Delta Dagger
(HAS Photo)

The Delta Dagger is a single-seat supersonic all-weather interceptor that is powered by one 17,200-lb after-burning thrust Pratt & Whitney J57-P-23 or -25 turbojet. It had a maximum speed of 825 mph, a service ceiling of 36,000', and a range of 1,350 miles. It was armed with two AIM-26 or 26A Falcon missiles, or one Falcon and two AIM-4C/D missiles in a weapons bay. *(The Complete Encyclopedia of World Aircraft).*

The primary mission of the F-102 was to intercept and destroy enemy aircraft. It was the world's first supersonic all-weather jet interceptor and the USAF's first operational delta-wing aircraft. The F-102 made its initial flight on 24 October 1953 and became operational with the Air Defense Command in 1956. At the peak of deployment in the late 1950's, F-102s equipped more than 25 ADC squadrons. Convair built 1,101 F-102s, 975 of which were F-102As. The USAF also bought 111 TF-102s as combat trainers with side-by-side seating. In a wartime situation, after electronic equipment on board the F-102 had located the enemy aircraft, the F-102's radar would guide it into position for attack. At the proper moment, the electronic fire control system would automatically fire the F-102's air-to-air rockets and missiles. *(Information courtesy of the USAF Museum).*

The F-102 shown above is on display at Peterson AFB near the Base Accommodations complex. Another may be viewed in the WORA&SM.

Convair F-106A Delta Dart
(HAS Photo)

The Delta Dart is a single-seat supersonic all-weather interceptor powered by one 24,500-lb afterburning thrust Pratt & Whitney J75-P-17 turbojet. It has a maximum speed of 1,525 mph, a service ceiling of 57,000', and a combat radius of 729 miles. *(The Complete Encyclopedia of World Aircraft).*

The Dart was first introduced to the USAF in 1960 to counter the threat of a nuclear attack by supersonic Russian bombers. The Delta Dart cruised at a speed near Mach 2, and its fire-control system worked with a ground-based radar to guide the aircraft and its weapons towards its target.

The delta wing gave the F-106 both stability and low drag for Mach 2 flight. The F-106 all-weather interceptor was developed from the Convair F-102 Delta Dagger. Originally designated the F-102B, it was redesignated F-106 because it had extensive structural changes and a more powerful engine. The first F-106A flew on 26 December 1956, and deliveries to the Air Force began in July 1959. Production ended in late 1960 after 277 F-106As and 63 F-106Bs had been built. The F-106 uses a Hughes MA-1 electronic guidance and fire control system. After takeoff, the MA-1 can be given control of the aircraft to fly it to the proper altitude and attack position. Then it can fire the Genie and Falcon missiles, break off the attack run, and return the aircraft to the vicinity of its base. The pilot takes control again for the landing. *(Information courtesy of the USAF Museum).*

The F-106 shown above is on display in the PA&SM Air Park. It was the 323rd F-106 produced, and was delivered to the USAF on 23 November 1960. It served with the 48th FIS at McChord AFB, Washington, but was damaged when the drop tanks were accidentally ejected on the ground. The resulting fire damaged both wings, the fuselage and vertical stabilizer. It was then sent to Lowry AFB in Colorado where it was used by the Technical Training Center from October 1975 to 02 November 1995, when it was trucked to Peterson AFB for display.

In order to reduce costs of restoring the F-106 at the museum, the 144th Fighter Wing of the Fresno, California Air National Guard adopted the aircraft. In recognition of their work, the aircraft is painted with the California ANG white tail and golden bear. It was formally dedicated on 9 September 2002. *(Information courtesy of the PA&SM).*

Convair HC-131A Samaritan
(HAS Photo)

The Samaritan is a twin-engine transport aircraft powered by a pair of 2,500-hp Pratt & Whitney R-2800-CB16 or CB17 radial piston engines. It had a maximum speed of 300 mph, a service ceiling of 24,900', and a range of 470 miles. *(The Complete Encyclopedia of World Aircraft)*.

The Samaritan was first built March 1954. Its primary mission was air evacuation, and the Coast Guard has used it extensively. It was also used by military as a flying classroom. The Samaritan was a USAF transport version of the Convair 240/340/440 series commercial airliners. The first Samaritan, a C-131A derived from the Convair 240, was delivered to the Air Force in 1954. It was similar to the T-29 trainer (also based on the Convair 240) flown by the USAF since 1949 to instruct navigators, bombardiers and radio operators. The C-131 was acquired primarily for medical evacuation and personnel transportation. While some T-29s also saw duty as staff transports, a few C-131s likewise were used for training and testing. In fact, the first prototype of the Southeast Asia vintage side-firing Gunship program used the C-131 airframe. Nearly all of the USAF's C-131s left the active inventory in the late 1970s, but a few were still serving in Air National Guard units in the mid-1980s. *(Information courtesy of the USAF Museum).*

The Navy's C-131 (R4Y) aircraft is the military version of Convair's 340 model commercial airliner which in turn is a newer design of the 240 model with a greater wing-span, wing area and weight; longer fuselage and a more powerful engine. First flown on 5 October 1951, the Convair 340 proved quite successful in the commercial field with sixteen American and foreign airlines placing orders for 160 production models in 1952 alone.

Delivery of 36 R4Y-1 aircraft (later redesignated C-131 in 1962) to the Navy began in 1952 and they were assigned to Navy and Marine Corps fleet support squadrons as logistic and administrative support aircraft. Capable of carrying up to 44 passengers and a freight payload of 12,000-lbs, the C-131 could also be employed as an aerial ambulance with a 27-bed capacity. A single R4Y-1Z had a VIP interior with 24 seats and became the VC-131F. Another was equipped with an extensive array of radar and electronic equipment (including radomes) for use as a countermeasures research and development aircraft. *(Information courtesy of the National Museum of Naval Aviation).*

The Pueblo Air Museum's aircraft (shown above) was used by the Coast Guard and was flown to the museum from the Colorado Surplus Property Agency from Davis Monthan AFB AZ. *(Information courtesy of the PWAM).*

Curtiss JN-4D Jenny

The Jenny was the standard two-seat pilot trainer of WWI. It is powered by a single liquid-cooled 90-hp V-8 90 HP OX-5 inline piston engine. It had a maximum speed of 75 mph, and a service ceiling of 6,500'. *(The Complete Encyclopedia of World Aircraft).*

The first airmail was flown in a Jenny between New York and Washington, D.C. on the 15th of May 1918. It was also the mainstay of the US Signal Corps and the barnstormers of the 1920's and early 1930's. One of America's most famous airplanes, the Jenny was developed by combining the best features of the Curtiss J and N models. A 1915 version, the JN-3, was used in 1916 during Pershing's Punitive Expedition into Mexico. Its poor performance, however, made it unsuited for field operations.

The JN-3 was modified in 1916 to improve its performance and redesignated the JN-4. With America's entry into World War I on April 6, 1917, the Signal Corps began ordering large quantities of JN-4s, and by the time production was terminated after the Armistice, more than 6,000 had been delivered, the majority of them JN-4D.

The Jenny was generally used for primary flight training, but some were equipped with machine guns and bomb racks for advanced training. After World War I, hundreds were sold on the civilian market. The airplane soon became the mainstay of the Barnstormer of the 1920s, and some Jennies were still being flown in the 1930s. *(Information courtesy of the USAF Museum).*

A Jenny was restored by the Antique Airplane Association of Colorado, and is presently located between gates B-54 and B-52 at Denver International Airport.

Curtiss P-40E Warhawk
(HAS Photo)

Curtiss continuously worked to improve the performance and effectiveness of the P-40. The E model was powered by one 1,150-hp Allison V-1710-39, which could maintain its power output to 11,700'. The armament varied, but the D version carried four and the E version was armed with six wing-mounted .50 cal machine-guns and the aircraft could carry one 500-lb of bomb or a 52-gal drop tank on an underfuselage hardpoint. These aircraft were provided to the Commonwealth forces with the designation Kittyhawk Mk. I. The P-40E-1 (Hawk 87-A3 and-A4) Kittyhawk served in virtually every theater of war. *(The Complete Encyclopedia of World Aircraft).*

The Curtiss P-40 fighter, which first flew in October 1938, came into being by modifying a radial engine P-36 Mohawk airframe to accept the new in-line Allison engine, resulting in a great saving of time and money. After its evaluation in May 1939, the US Army Air Corps, and the British and French governments placed large orders.

During its service life, it was produced in a number of models, with various armament, engine, and airframe modifications, with the P-40N being the most highly-produced version (over 5,000). The names Warhawk and Kittyhawk were also used in referring to various models of the P-40.

P-40s were in constant wartime service from Pearl Harbor until VJ Day, and fought in every theater, from every type of airfield, in every type of weather. Although they were not deemed suitable for use in Britain, because of their limited service ceiling, they served well with the U.S., British, Canadians, Australians, New Zealanders, French and Russians. They also served in places such as China, the African desert, the Aleutians, Iceland, the Southwest Pacific, the Philippines, on the Eastern Front and even flying off the CVE Chenango which had ferried them across the Atlantic for the invasion of North Africa! It was versatile, honest and rugged, but behind in comparisons with the best contemporary fighters. It was said that P-40 pilots expected to be out-classed and out-numbered, but they had such confidence in their plane's ability to get them home, that they never hesitated to engage the enemy. At least 85 pilots became Aces flying P-40s, including over 25 Flying Tigers, and not counting Russian pilots.

Naval Aviation's connection with the P-40 is through the American Volunteer Group (AVG), the famous Flying Tigers, the majority of whom (about 60%) were Navy or Marine Corps personnel. The AVG initially received 100 Curtiss Hawk H81A-3s, export version of the P-40C, which were side-tracked from a Lend-Lease order for Britain, and shipped to Burma. Instead of the C's armament (six wing mounted .30s), they had the same armament as the Bs (four .30s in the wings and two .50s firing through the propeller). Later the Flying Tigers received some P-40Es with six wing-mounted .50s. From the AVG's first action on 20 December 1941, until their replacement by the U.S. 10th Air Force 4 July 1942, they reportedly shot down over 286 enemy planes with the loss of only four pilots in aerial combat!

The P-40 had an interesting connection with the P-51 Mustang, considered by many the best fighter of World War II. The British had approached North American Aviation to manufacture P-40s for them under license, and North American proposed that they design a brand new fighter instead. The British accepted, and this new fighter became the P-51. *(Information courtesy of the National Museum of Naval Aviation).*

The original P-40N Warhawk at PAFB (Serial No. 42-105927), was sent to the Robins Museum of Aviation in Florida June 1994. A P-40E fiberglass replica as shown above has replaced it. It is on display mounted on a pylon West of the PA&SM Air Park.

De Havilland DHC-2/U-6A Beaver Mk. I

The U-6A Beaver is a light utility transport powered by one 450-hp Pratt & Whitney piston engine. It has a top speed of 163 mph, a service ceiling of 20,000', and a range of 455 miles. Although the Beaver was unarmed, it does have provision for two racks under each wing capable of carrying one 250 lb. bomb or chemical tank on each rack.

The U-6A (known as the L-20A until 1962), was manufactured by DeHavilland Aircraft of Canada, Ltd. Between 1952 and 1960, nearly 1,000 were delivered to the U.S. armed services. Most were used by the U.S. Army, but more than 200 U-6As went into the USAF inventory and were flown in utility transport and liaison roles.

The principal mission of the USAF U-6A was aerial evacuation of litter and ambulatory patients. Other missions included courier service, passenger transport, light cargo hauling, reconnaissance, rescue, and aerial photography. The U-6A saw USAF service in both the Korean and Vietnam Wars. *(Information courtesy of the USAF Museum).*

A number of Beavers are flown by private owners in Colorado.

Douglas B-18/DB-1 Bolo
(HAS Photo)

The Bolo is a twin-engine medium bomber and ASW aircraft powered by a pair of 1000-hp Wright R-1820-53 Cyclone 9-cylinder radial piston engines. It had a maximum speed of 346 km/h, a service ceiling of 23,900', and a range of 1,931 km. It was armed with three .3-in machine-guns, with one each in the nose, ventral and dorsal positions. The Bolo could also carry up to 6,500-lbs of bombs. *(Information courtesy of the Canadian Forces Archives)*

The Douglas Aircraft Co. developed the B-18 to replace the Martin B-10 as the Army Air Corps' standard bomber. The Bolo's design was based on the Douglas DC-2 commercial transport. During Air Corps bomber trials at Wright Field in 1935, the B-18 prototype competed with the Martin 146 (an improved B-10) and the four engine Boeing 299, forerunner of the B-17. Although many Air Corps officers believed the Boeing design was superior, only 13 YB-17s were initially ordered. Instead, the Army General Staff selected the less costly Bolo and, in January 1936, ordered 133 as B-18s. Later, 217 more were built as B-18As with a "shark" nose in which the bombardier's position was extended forward over the nose gunner's station. By 1939, under-powered and with inadequate defensive armament, the Bolo was the Air Corps' primary bomber. Some B-18s were destroyed by the Japanese on the 7th of December 1941. By early 1942, improved aircraft replaced the Bolo as a first-line bombardment aircraft. Many B-18's were then used as transports, or modified as B-18Bs for anti-submarine duty. *(Information courtesy of the USAF Museum)*.

The Bolo shown above is on display inside the WORA&SM.

Douglas C-47A Skytrain

The Skytrain is a short to medium range transport powered by a pair of 1,200-hp Pratt & Whitney R-1830-S1C3G Twin Wasp radial piston engines. It had a maximum speed of 230 mph, a service ceiling of 23,200', and a range of 2,125 miles. The Skytrain was also configured as a gunship, the AC-47D. The Spooky gunship variant was armed with three .3-in General Electric Miniguns firing through the fourth and fifth windows and from the open door on the port side of the fuse-lage. *(The Complete Encyclopedia of World Aircraft)*.

The Douglas C-47 cargo and passenger aircraft was the workhorse of the armed forces for many years. It carried heavy loads to high altitude and was used extensively in the China-Burma-India Theater of WWII to fly supplies over the hump (The Himalayan Mountains). It was used by paratroopers and figured prominently in the European D-Day invasion.

Few aircraft are as well known or were so widely used for so long as the C-47 or Gooney Bird as it was affectionately nicknamed. The aircraft was adapted from the DC-3 commercial airliner, which appeared in 1936. The first C-47s were ordered in 1940 and by the end of WW II, 9,348 had been procured for AAF use. They carried personnel and cargo, and in a combat role, towed troop-carrying gliders and dropped paratroops into enemy territory. After WW II, many C-47s remained in USAF service, participating in the Berlin Airlift and other peacetime activities. During the Korean War, C-47s hauled supplies, dropped paratroops, evacuated wounded and dropped flares for night bombing attacks. In Vietnam, the C-47 served again as a transport, but it was also used in a variety of other ways, which included flying ground attack (gunship), reconnaissance, and psychological warfare missions. *(Information courtesy of the USAF Museum).*

A C-47 is on display at the PWAM. This C-47 was last used by the military as an Arctic research laboratory out of Point Barrow Alaska. It was donated in 1979 by the US Naval Air Systems Command at Washington D.C. and was flown to Pueblo from Davis Monthan AFB AZ. *(Information courtesy of the PWAM).*

Douglas A-26C Invader
(HAS Photo)

The Invader was the fastest three-seat USAAF light/medium bomber of WWII. A pair of 2,000-hp Pratt & Whitney R-2800-27 or -79 Double Wasp radial piston engines powered it. It had a maximum speed of 355 mph, a service ceiling of 22,100', and a range of 1,400 miles. It was armed with 10 .50 cal machine-guns and could carry up to 4,000-lbs of bombs. *(The Complete Encyclopedia of World Aircraft)*.

The initial A-26 was a follow-up airplane to the A-20 Havoc, and made its first flight on 10 July 1942. Production delivery began in August 1943, and on November 19, 1944, it went into combat over Europe. It was used for level bombing, ground strafing and rocket attacks. The first deliveries of the A-26B Invader began in April 1944, with 1,355 being produced. Some variants of the Invader such as the C model have a glazed nose for the bombardier, while other Attack variants were equipped with up to eight .50 caliber guns in the nose for the tactical role. By the time production halted after VJ-Day, 2,502 Invaders had been built.

The A-26 was redesignated the B-26 in 1948. During the Korean War, the airplane entered combat once again, this time as a night intruder to harass North Korean supply lines. Early in the Vietnam conflict, the Invader went into action for the third time. Also, the USAF ordered 40 modified B-26Bs having more powerful engines and increased structural strength. Designated the B-26K, the airplanes were designed for special air warfare missions. In 1966, the B-26K was redesignated the A-26A. The TB-26B Invader was a training variant of the B-26; the RB-26C was a night reconnaissance variant. *(Information courtesy of the USAF Museum).*

The PWAM's A-26C shown above was purchased by the city of Pueblo, in May 1972. *(Information courtesy of the PWAM).*

Douglas F-6A Skyray
(HAS Photo)

The Skyray was the first short range, fast climbing, carrier-based interceptor. The Skyray is powered by one 14,500-lb thrust Pratt & Whitney J57-P-8B turbojet. It has a maximum speed of 7722 mph at sea level, a service ceiling of 55,000', and a range of 1,200 miles. It is armed with four fixed forward-firing 22mm cannon and can carry up to 4,000-lbs of stores, including auxiliary fuel, bombs, rockets or missiles on six underwing hardpoints. *(The Complete Encyclopedia of World Aircraft).*

When the USAF accepted the Navy F4D into its domain its inventory it already had an F4 fighter so it designated the aircraft as a model F6A. Extensive wind tunnel tests led to the creation of the bat-like form with the horizontal flying controls attached to the wing trailing edge classifying it as tailless. It set a world speed record of 753.4 mph for a carrier plane.

The Skyray was one of the most effective interceptors of its era. In June 1947 Douglas Aircraft Corporation received a Navy contract for the study of a delta-wing fighter. Approval of preliminary designs and engineering concepts 18 months later led to a contract for two prototypes (XF4D-1s) that were delivered and first flown in January 1951. The XF4D-1 proved not to be a true delta-wing but rather a swept-wing with low aspect ratio. Testing trials followed by carrier suitability tests proved quite successful and full-scale production of 420 F4D Skyrays commenced thereafter. During test phases the XF4D-1 prototypes established speed records over the International 3-km course (755 mph) and the 100 km closed circuit course (728 mph).

Armed with 20-mm cannons and Sidewinder air-to-air missiles, the F4D was capable of carrying a weapons load compatible with its mission to intercept enemy aircraft before they reached their target. As an interceptor, the F4D established five world rate-of-climb records, which, in turn, led to the assignment of a Navy all-weather F4D squadron at San Diego, and one at Key West to the Air Force's North American Air Defense Command in an interceptor role. The San Diego unit earned honors as the best in NORAD for two years running. The last of the short-ranged Skyrays served until 1964. *(Information courtesy of the National Museum of Naval Aviation).*

The Pueblo Air Museum's F-6A shown above was used at the Emily Griffith Aircraft Training Facility and was donated to the museum by

The Colorado Supply Program Agency. *(Information courtesy of the PWAM).*

Fairchild C-119B Flying Boxcar
(HAS Photo)

The C-119 is a twin-tailed short-nosed transport aircraft converted to a gunship. The Boxcar was first produced in 1949 as an Assault Transport aircraft. A pair of 3,500-hp Wright R-3350-85 Duplex Cyclone 18-cylinder radial piston engines powered the Flying Boxcar. It had a maximum speed of 200 mph, and a range of 2,280 miles. *(The Complete Encyclopedia of World Aircraft).*

The C-119, developed from the WW II Fairchild C-82, was designed to carry cargo, personnel, litter patients, and mechanized equipment, and to drop cargo and troops by parachute. The first C-119 made its initial flight in November 1947 and by the time production ceased in 1955, more than 1,100 C-119s had been built. The USAF used the airplane extensively during the Korean War and many were supplied to the U.S. Navy and Marine Corps and to the Air Forces of Canada, Belgium, Italy, and India. In South Vietnam, the airplane once again entered combat, this time in a ground support role as the AC-119G gunship. The AC-119G gunships mounted four side-firing .3-in six-barrel Miniguns capable of firing up to 6,000 rounds per minute per gun. The gunship was equipped with armor protection and flare-launchers. *(Information courtesy of the USAF Museum).*

The PWA Museum's C-119 shown above was donated in 1977 by The US Naval Air Systems command in Washington D.C. and flown to Pueblo from Davis Monthan AFB AZ. *(Information courtesy of the PWAM).*

Fairchild Republic A-10A Thunderbolt II
(HAS Photo)

The Thunderbolt is a single-seat close-support aircraft powered by a pair of 9,065-lb thrust General Electric TF34-GE-100 turbofans. It has a maximum speed of 439 mph, and an endurance rate of 1 hour and 40 minutes. The Thunderbolt II is armed with a General Electric GAU-8/A Avenger 30mm seven-barrel cannon. *(The Complete Encyclopedia of World Aircraft).*

The A-10 is the first USAF aircraft designed specifically for close air support of ground forces. It is named for the famous P-47 Thunderbolt, a fighter often used in a close air support role during the latter part of WW II. The A-10 is designed for maneuverability at low speeds and low altitudes for accurate weapons delivery, and carries systems and armor to permit it to survive in this environment. It is intended for use against all ground targets, but specifically tanks and other armored vehicles. The Thunderbolt II's great endurance gives it a large combat radius and/or long loiter time in a battle area. Its short takeoff and landing capability permits operation from airstrips close to the front lines. Service at forward area bases with limited facilities is possible because of the A-10's simplicity of design. The first prototype Thunderbolt II made its initial flight on May 10, 1972. A-10A production commenced in 1975. Delivery of aircraft to USAF units began in 1976 and ended in 1984. *(Information courtesy of the USAF Museum)*.

The A-10 shown above is the most recent static display aircraft to the USAF Academy, where it is mounted on a pylon beside the Northrop T-38 Talon. This particular aircraft has an interesting history, in that it is the only A-10 to have recorded an official air-to-air kill against an Iraqi helicopter. The Iraqi combatant was destroyed in flight when a bomb dropped by this A-10 went off during an attack in "Operation Desert Storm." *(Information courtesy of the USAF Academy)*

Fokker D.VII

The Fokker D.VII is a single-seat fighter/scout powered by one 185-hp BMW III 6-cylinder inline piston engine. It has a maximum speed of 124 mph, a service ceiling of 22,965', and an endurance rate of 1½ hours. It is armed with two fixed forward-firing 7.92mm LMG 08/15 machine-guns. *(The Complete Encyclopedia of World Aircraft)*.

Historically, the Fokker D.VII is considered by most authorities to be the most significant combat aircraft of World War I. Created by the great Fokker engineering genius Reinhold Platz, the D.VII was the scourge of the skies over Europe during the last year of the war. Pilots found the D.VII easy to fly, confidence-generating and with specific control attributes that permitted distinct advantages when engaged in air-to-air combat (stalls, for instance, were predictable and controllable). *(Information courtesy of the Museum of Flight)*.

First appearing over the World War I battlefield in May 1918, the Fokker D.VII quickly showed its superior performance over Allied fighters. With its high rate of climb, higher ceiling, and excellent handling characteristics, the German pilots were able to score 565 victories over Allied aircraft during August 1918. Designed by Reinhold Platz, the D.VII was chosen over several other designs during a competition held in January and February 1918. Baron Manfred von Richthofen, the famous Red Baron, flew the prototype, designated VII. He found it easy to fly, able to dive at high speed quickly yet remain steady as a rock, and had good visibility for the pilot. His recommendation virtually decided the competition. To achieve higher production rates, Fokker, the Albatross company, and the Allgemeine Elektrizitats Gesellschaft (A.E.G.) all built the D.VII. By war's end in November 1918, these three companies had built more than 1,700 aircraft. *(Information courtesy of the USAF Museum)*.

A Fokker D.VII replica can be viewed at the Platte Valley Air Park.

Fokker Dr.I Dreidecker

The Fokker Dr.I is a single-seat fighter and scout Triplane powered by one 110-hp Thulin-built Le Rhône or Oberursel Ur.II 9-cylinder rotary piston engine. It has a maximum speed of 103-mph at 13,125', a service ceiling of 20,000', and an endurance rate of 1 1/2 hours. It is armed with two fixed forward-firing 7.92 (.31-in) LMG 08/15 machine-guns. *(The Complete Encyclopedia of World Aircraft)*.

The Fokker Dr.I is a full-scale replica of one of the most famous German combat aircraft of World War I. Many competent and well-known German pilots flew the Dr.I Dreidecker, often referred to informally as the Fokker Triplane. The most famous of these was the legendary Manfred von Richthofen, who first flew a Dreidecker during September of 1917. He remained a strong proponent of the type until his death in Dr.I Serial No. 425/17 on 21 April 1918.

There are no known surviving original Fokker Dr.I's. The last original aircraft was destroyed during an Allied bombing raid on Berlin during World War II. *(Information courtesy of the Museum of Flight)*.

A Fokker Dr.I can be viewed at the Platte Valley Air Park.

Folland Fo.141 Gnat T Mk. 1

The Gnat is a two-seat advanced trainer powered by one 4,230-lb thrust Bristol Siddeley Orpheus 100 turbojet. It has a maximum speed of 636 mph, a service ceiling of 31,000', and a range with two underwing tanks of 1,151 miles. The Gnat T Mk. 1 was the last production variant, first flown in the UK on 9 April 1965. It was flown by the British Red Arrows aerial demonstration team. *(The Complete Encyclopedia of World Aircraft)*.

General Dynamics FB-111A Aardvark

The F-111 is a two-seat long-range, multi-purpose all-weather strike aircraft capable of navigating at low level to reach targets deep in enemy territory and to deliver ordnance on the target. Primarily a bomber, the F-111 featured a sweep wing varying between 16 degrees and 72.5 degrees, with side-by-side seating for a pilot and weapons systems officer.

Primarily a bomber, the F-111 featured a sweep wing varying between 16 degrees and 72.5 degrees, with side-by-side seating for a pilot and weapons systems officer. The versatile swing-wing F-111, unofficially named the Aardvark until its retirement ceremony on July 27, 1996 where the name was made official, entered the USAF inventory in 1967. The F-111's wings are straight for take-offs and landings, or slow speed flight; by sweeping its wings rearward it can exceed twice the speed of sound (Mach 2). The F-111 was a long-range, all-weather strike aircraft capable of navigating at low level to reach targets deep in enemy territory and to deliver ordnance on the target. It was developed in 1960 when the Department of Defense combined the USAF's requirement for a fighter-bomber with a Navy need for an air-superiority fighter, though the Navy eventually cancelled its program.

The F-111A was the initial production version powered by a pair of 18,500-lb thrust Pratt & Whitney TF-P-3 turbofan engines. 158 of them were built.

The EF-111A Raven is an electronic counter-measures (ECM) tactical jamming version produced by Grumman as a conversion from the F-111A.

The FB-111A is a two-seat strategic bomber version made for USAF Strategic Air Command. This variant had a 7' increase in wingspan, strengthened landing gear, an increased fuel capacity, and was powered by a pair of 20,150-lb thrust TF30-P-7 engines. 76 were built.

The RF-111A is a reconnaissance conversion of the F-111A.

The F-111B is a carrier-based fleet defense aircraft manufactured for the US Navy. 7 were built.

F-111D similar to the F-111E but powered by a pair of 19,600-lb thrust Pratt & Whitney TF30-P-9 turbofan engines. This version featured an improved Mark II avionics package to enhance its air-to-air weapon deployment and navigational capability, and an environmental control system. 96 were built.

The F-111E superceded the F-111A on the production line from the 160th aircraft. It has modified air intakes and is powered by two 25,100-lb thrust Pratt & Whitney TF-30-P-3 augmented turbofan engines. It has a maximum speed of 1,650 mph, a service ceiling of 59,000', and a range of 2,925 miles. It is armed with on 20mm multi-barreled M61A-1 cannon and carries on 750-lb B43 bomb or two B43 bombs in an internal weapons bay. It is also equipped with three underwing hardpoints on each outer wing panel, with the inner four pivoting to keep stores aligned as the wings sweep. 94 were built.

The F-111F model was equipped with an all-weather AN/AVQ-26 Pave Tack infrared targeting designator/reader carried in a pod-mounted turret. It could track and designate ground targets for targets for laser, infrared and electro-optical bombs. In all, 562 F-111s of all series were built. *(Information courtesy of the USAF Museum).*

An FB-111A is on display inside the WORA&SM.

General Dynamics F-16A Fighting Falcon

The Fighting Falcon is a single-seat front-line air-superiority and air-to-ground attack fighter, powered by a Pratt & Whitney F100-PW-220 engine, or by a General Electric F110-GE-100 after-burning turbofan. It has a maximum speed of more than 1,320 mph, a service ceiling above 50,000', and a range of 575 miles. *(The Complete Encyclopedia of World Aircraft).*

In use with the Thunderbirds, the F-16 Fighting Falcon is one of the U.S. Air Force's frontline jet fighter aircraft. Since its introduction in 1976, the F-16 has made its mark as a true multi-role fighter. Its highly maneuverable design has proven itself in air-to-air combat and air-to-ground attack. This aircraft provides a relatively low-cost, high-performance weapons system for the air forces of the free world.

In an air combat role, the F-16's maneuverability and combat radius (distance it can enter air combat, stay, fight, and return) exceed that of all potential threat fighter aircraft. It can locate targets in all weather conditions and detect low flying aircraft in radar ground clutter. In the air-to-ground role, the F-16 can fly more than 500 miles, deliver its weapons with superior accuracy, defend itself against enemy aircraft, and return to its origin base. An all-weather capability allows it to accurately deliver ordnance during non-visual bombing conditions.

In designing the F-16, advanced aerospace technology and proven reliable systems from other aircraft were selected. These were combined to simplify the jet and reduce its size, weight, purchase price, and maintenance costs. The light weight of the fuselage is achieved without reducing its strength. The F-16 can withstand up to nine G's (nine times the force of gravity) with its internal fuel tanks filled, greater than any other current fighter aircraft. The cockpit and its bubble canopy are designed to give the pilot unobstructed forward and upward vision, and greatly improved vision over the side and to the rear. The seat-back angle was expanded from a standard 13 degrees to 30 degrees, increasing pilot comfort and G-force tolerance. The pilot has excellent flight control of the F-16 through its "fly-by-wire" system. Electrical wires relay commands, replacing the cables and linkage controls. For easy and accurate control of the aircraft during high G-force combat maneuvers, a side-stick controller is used instead of the conventional center-mounted stick. Hand pressure on the side-stick controller sends electrical signals to actuators of flight control surfaces such as ailerons and rudder.

Avionics systems include a highly accurate inertial navigation system in which a computer provides global positioning and steering informa-tion to the pilot. The plane uses UHF and VHF radios plus an instru-ment landing system. It also has a radio/threat warning system and modular counter measure pods to be used against airborne or surface electronic threats. The fuselage also has space for additional avionics systems.

All F-16s delivered since November 1981 have built-in structural and wiring provisions and systems architecture that permits expansion of the jet's multi-role flexibility to perform precision strike, night-attack and beyond-visual-range interception missions. This improve-ment program led to the F-16C/D aircraft, which are the single and two-place counterparts to the F-16A/B, respectively, and incorporate the latest cockpit control and display technology.

Now serving at U.S. Air Force bases worldwide, the Air National Guard and Air Force Reserve also fly the Fighting Falcon. In addition, the F-16 is a vital part of the air forces of Belgium, Denmark, Nether-lands, Norway, Israel, Egypt, Korea, Pakistan, Venezuela, Turkey, Greece, Singapore, Thailand, Indonesia, and Bahrain. The worldwide use of the F-16 truly makes it a mainstay of international tactical air forces. *(Information courtesy of the Public Affairs office of the USAF Thunderbirds).*

A Falcon is on display at Buckley AFB. Another is displayed at on the grounds below the Chapel at the USAF Academy in Colorado Springs. The aircraft on display at the USAFA was a non-flying test bed aircraft and it is painted in the colours of the 57th Fighter Weapons Wing located at Nellis AFB, Nevada. It was acquired by the Academy in 1982, and is maintained by the "3rd Cadet Group."

Gnome and Le Rhône Rotary Engine

Gnome and Le Rhône engines were the earliest radial rotary engines built. These engines played a big part in WWI. The earlier Gnome was replaced by Le Rhône engines of similar design, which employed a slightly different method of connecting rod attachment to the crankshaft. Some also employed only a single push rod for both valves. The unique feature of all of these engines was that the crankshaft remained stationary. The crankshaft was rigidly attached to the airframe and the prop, cylinders and crankcase rotated as one piece. Fuel was fed through the hollow crankshaft into the crankcase. In early models only a spray nozzle was used, past which the air flowed on its way into the crankcase. The later Le Rhône did use some induction pipes. Gnome Valves consisted of an intake in the cylinder head and an exhaust valve in the top of the cylinder. Later models—Le Rhône—eliminated one valve, using instead a port drilled in the cylinder skirt. It retained the head valve and was frequently referred to as the "Monosoupape."

Gnomes were built in various sizes: 7, 9, & 14 cylinder. The latter developed 180 HP. Neither engine had a throttle and engine control was ignition "on off" controlled by a button atop the control stick called the "coupe" button. The PWAM's Gnome engine was purchased for the Weisbrod Aircraft Museum and donated by the Pueblo Historical Aircraft Society. The PWAM's engine was completely disassembled and restored by the PHAS team of: Ted Baer, Gerald Putnam, Joe Harman, Bob Dyleski and R. K. Darr. It was so badly rusted that it had to be soaked for weeks in various rust-removal solutions before the pistons could be forced out of the sleeves and then only by wielding of appropriately sized hammers and extreme pressure applied by various hydraulic means. *(Information courtesy of the PWAM)*.

Goodyear FG-1D Corsair
(HAS Photo)

The FG-1D/F4U Corsair is a single-seat carrier-based fighter powered by one 2,000-hp Pratt & Whitney R-2800-8 radial piston engine. It has a maximum speed of 417 mph, a service ceiling of 36,900', and a range of 1,015 miles. It is armed with six wing-mounted .50 cal machine-guns. *(The Complete Encyclopedia of World Aircraft)*.

The Corsair is equipped with a radial air-cooled engine mounted in the nose and fitted with a four-blade propeller. Its low-set wings were inversed-gullwing shaped. They were elliptical with rounded tips. An oval air intake was housed in the leading edge of either wing close to the fuselage. The tailplane was parabola-shaped and was set past the tail fin. The Corsair's landing gear with tail wheel and a hook was retractable.

The Corsair first flew on 29 May 1940 and was built by three different companies, Chance Vought (F4U); Goodyear (FG-1); and Brewster (F3A). The Corsair was in production longer than any other World War Two fighters were. The F4U started production in June 1942 with the last Corsair rolling off the assembly line in December 1952 for a total production run of 12,571 Corsairs built. There were 88 Marine Aces in 20 Marine squadrons flying the F4U Corsair during World War Two.

The Corsair was conceived to use the most powerful engine and largest propeller a fighter plane ever had. Unfortunately, the long propeller blades caused ground clearance problems. The W or gull-shaped-wings helped to reduce the required length of the landing gear, and provided quite good visibility to the sides. The landing gear, however, was still long and therefore fragile, and the first reports of carrier tests were: *"This plane will never be a good carrier plane."* The Corsair was nicknamed the Ensign Eliminator. It would appear that the strong engine torque didn't make it an easy aircraft to land.

The F4U was developed in 1938 for the US Marine Corps, with the first prototype XF4U-1 flying on the 29th of May 1940. The airplane went in production in 1942. The main modifications were: F4U-1—the first production version; F4U-1A—an improved F4U-1; F4U-1B—exported to United Kingdom in 1943; F4U-1C—a close air support version armed with guns rather than machine guns; F4U-1D—a fighter/bomber; F4U-2—a night fighter, with a radar which was housed in a fairing on the left wing; XF4U-3—an experimental high-altitude fighter with twin turbochargers; F4U-4—fitted with a more powerful engine, and which could carry rockets; F4U-5—a night fighter with a Pratt & Whitney R-2800 engine. By 1952, 12,681 Corsairs had been built. Many F4U Corsairs were used in Korea. *(Information courtesy of the National Museum of Naval Aviation)*.

The F4U-5 Corsair, too late for World War Two, started coming off the assembly line in April 1946. The night fighting units flying the F4U-5N Corsair, F7F-3N Tigercat and the F3D Skyknight accounted for 11 enemy aircraft during the Korean War. Sept 10th 1952 Capt. Jesse G. Folmar of VMF-312 is credited with shooting down a Russian-made MIG-15 while flying the Corsair. *(Information courtesy of the Flying Leathernecks Aviation Museum).*

The FG-1D shown above is a "Super Corsair" racing aircraft on display at the WORA&SM. It has had 3 ½ feet clipped off each wing, and has been modified and equipped with a Buick engine. *(Information courtesy of the WORA&SM).*

Grumman G-44A/J4F-1 Widgeon

The Widgeon is a five-seat light transport amphibian aircraft powered by a pair of 200-hp Ranger L-440C-5 6-cylinder piston engines. It has a maximum speed of 153 mph, a service ceiling of 14,600', and a range of 920 miles. *(The Complete Encyclopedia of World Aircraft).*

The Coast Guard obtained 25 Grumman G-44 light amphibian aircraft during 1941-42 as J4F-1s, later adding light bomb racks under the wings for service with USCG patrol squadrons. One hundred thirty one J4F-2s subsequently procured by the Navy, carried a crew of two and three passengers.

For many years it was believed that a Coast Guard J4F Widgeon flown by Aviation Chief Machinist's Mate Henry White and Radioman First Class George H. Boggs, Jr., sank the German submarine U-166 south of Houma, Louisiana on 1 August 1942. However, in June 2001 oil companies scouting the route for a proposed gas pipeline in the Gulf of Mexico discovered the wreckage of the U-166 near that of SS Robert E. Lee, which the U-boat sank on 30 July 1942. According to World War II action reports, Robert E. Lee's escort, PC-566, attacked a submarine that day with inconclusive results. The discovery of U-166 appears to indicate that PC-566's attack was successful and that the Coast Guard J4F attacked one of two other U-boats operating in the area, neither of which was sunk. *(Information courtesy of the National Museum of Naval Aviation).*

Grumman TBM-3E Avenger

The Avenger is a three-seat carrier-based torpedo-bomber. The TBM-3 variant is powered by one 1,750-hp Wright R-2600-20 Cyclone 14-cylinder radial engine. It had a maximum speed of 267 mph, a service ceiling of 23,400', and a range of 1,130 miles. It was armed with two forward-firing .50 cal machine-guns, one .50 cal machine-gun in a dorsal turret and on .3-in machine-gun in a ventral position. It could carry 2,000-lbs of weapons in the bomb bay, and later variants could be armed with rocket projectiles, drop-tanks or be equipped with a radar pod mounted under the wing. *(The Complete Encyclopedia of World Aircraft).*

The Avenger was the workhorse of the USN during WWII. Avengers remained in service until the mid-1950s. They served in virtually every carrier and was employed for torpedo attack, level bombing and spotting/reconnaissance. Many are used in the water-bomber role today.

In response to a 1939 requirement for a carrier-based torpedo-bomber to replace the TBD Devastator, Grumman Aircraft Engineering Corporation produced the TBF Avenger equipped with an electrically powered gun turret and an internal bomb-bay for four 500-lb bombs or an aerial torpedo. Crewed by a pilot, radioman and gunner, the TBF could cruise at 145 mph. In 1942 General Motors also built the TBF known as the TBM. The Eastern Aircraft Division of General Motors began producing the Avenger in 1943 while Grumman refitted its production line to build other aircraft. A total of 9,842 TBF/TBM Avengers were produced, the last of which were retired in 1954.

The combat debut of the TBF occurred during the Battle of Midway when six of them were launched from Midway without fighter escort to attack the approaching Japanese fleet. Jumped by Zeros during their torpedo runs, five were shot down while the sixth was so badly shot-up (pilot/radioman wounded, gunner killed) that it was scrapped after landing at Midway with bomb-bay doors open and only one wheel extended.

In 1944 while flying from an aircraft carrier on a bombing mission against a Japanese radio station on Chichi Jima, the TBM of LTJG George Bush was severely damaged by anti-aircraft fire. Despite a flaming engine, he continued his dive to score a direct hit before being forced to bail-out over water where he was picked up by a submarine and subsequently returned to his squadron to fly additional combat missions. Both of his crewmen failed to survive.

Other combat highlights involving the Avenger included the sinking of six Japanese transports in one day by aircraft from a single squadron; the sinking of the world's largest battleships, the Musashi and Yamato; shooting down ninety-eight planes in aerial combat; and sinking thirty-one German submarines in the Atlantic.

In 1945, five TBMs from NAS Ft. Lauderdale, Florida, on a routine training flight disappeared in an area known as the Bermuda Triangle. Their location remains a mystery to this day. *(Information courtesy of the National Museum of Naval Aviation).*

An Avenger flies with the CAF at Grand Junction.

Grumman HU-16C Albatross

The Albatross is a twin-engine general-purpose amphibian aircraft with good short take-off and landing characteristics and a range of 4600 km. It could also carry 10 passengers or 12 stretcher patients. A pair of 1,425-hp Wright R-1820-76A or 76B Cyclone 9-cylinder radial piston engines powered the Albatross. It has a maximum speed of 236 mph, a service ceiling of 21,500', and a range of 2,850 miles. *(The Complete Encyclopedia of World Aircraft).*

The versatile Albatross amphibian was designed to meet a Navy requirement for a utility aircraft, which could operate from land or water and, with skis, from snow and ice. The prototype first flew on October 24, 1947 and soon after the USAF ordered a quantity for air-sea rescue duties as SA-16As. (In 1962 the USAF designation was changed to HU-16.) Grumman delivered 297 -As to the Air Force; most were assigned to the Air Rescue Service.

In 1955, Grumman developed an improved version with a 16-½ foot increase in wing-span and larger aileron and tail surfaces. Beginning in 1957, many -As were converted to the -B configuration with these improvements.

The Albatross is best known as a rescue aircraft. During the Korean War, Albatrosses rescued almost 1,000 United Nations personnel from coastal waters and rivers, often behind enemy lines. They also made numerous dramatic and hazardous rescues in Southeast Asia, on occasion taxiing many miles over rough, open water when unable to take-off. *(Information courtesy of the USAF Museum).*

Begun in 1944 as a general-purpose amphibian to replace the Grumman JRF Goose, which served throughout WW II, the HU-16 (UF) Albatross was another example of Grumman's ten-year experience in building amphibians. Flown in 1947, it featured a conventional two-step hull, full amphibian capability, a high wing with fixed stabilizing floats, and a single tail unit. Three variations of the aircraft were produced, including the winterized UF-1L for Antarctic operations, the dual-control UF-1T for use as a trainer, and the general purpose UF-1. The U.S. Air Force also procured the Albatross as the SA-16 for air-sea rescue duties, and signed an agreement in which the Air Force and Coast Guard would train Navy pilots in the techniques of operating the HU-16 in Search and Rescue.

The Albatross established three record flights for amphibians in September 1962 that included two separate altitude flights carrying 1,000 and 2,000-kg loads to 29,460' and 27,380' respectively, and a new world 5,000 km. speed record carrying a 1,000-kg load with a speed of 151.4 mph in a UF-2G. The last operational HU-16 Albatross made a final water landing on Pensacola Bay on 8 August 1976 and was then turned over to the NMNA. *(Information courtesy of the National Museum of Naval Aviation).*

Grumman F9F-8 Cougar
(HAS Photo)

The Cougar was a faster, swept-wing version of the USN's first jet carrier-based fighter, the Panther. It was powered by one 7,200-lb thrust Pratt & Whitney J48-P-8A turbojet and had a maximum speed of 705 mph, a service ceiling of 50,000', and a range of 600 miles. The Cougar was armed with two 20-mm cannon and could carry up to 2,000-lbs of weapons on underwing hardpoints. *(The Complete Encyclopedia of World Aircraft)*.

The Cougar came into service in 1952, after the F9 first made its appearance in 1951, as an outgrowth of the Grumman Panther. At that time it was called the XF9F-6, the first of the Panther series with a swept back wing. The next in the series was the F9F7, a lower thrust version used solely as a day fighter by the US Marine Corps. Then came the F9F-8, later re-designated as F9 and modified for a variety of uses including photo-reconnaissance and target drones. The F9-8 was the Blue Angels first swept wing and was used from 1955-1958. The first production model flew on Jan. 18, 1954 and went into service in Oct 1954. The F9-6 version set an unofficial transcontinental speed record of 3 hours, 45 minutes, 20 seconds on 1 April 1954. The nose boom on these aircraft was used for in-flight re-fueling. *(Information courtesy of the National Museum of Naval Aviation)*.

The Aurora CO Chamber of Commerce donated the newly painted F-8 shown above to the Pueblo Weisbrod Air Museum. *(Information courtesy of the PWAM)*.

Grumman F11F-1 Tiger
(HAS Photo)

The Tiger is a single-seat shipboard fighter-bomber powered by one 7,450-lb thrust Wright J65-W-18 turbojet. It had a maximum speed of 750 mph at sea level, a service ceiling of 41,900', and a range of 1,270 miles. It is armed with four 20mm cannon and four underwing Sidewinder air-to-air missiles. *(The Complete Encyclopedia of World Aircraft)*.

Originally the F9F-9 was to have evolved out of the earlier Panther/ Cougar aircraft achieved an entirely new design and the F-9F9 was designated as the F11A Tiger. It went supersonic easily and served with the US fleet as a carrier-based attack and fighter plane. One of the most unusual features was use of spoilers instead of ailerons for bank & roll. This permitted flaps, which extended the full length of the trailing edge of the wing. Two Tigers were fitted with the famous J79 engine of

15,000 pounds thrust and achieved a high speed of 1200 mph & climb of 79,938'.

Designed as a lightweight, air-superiority fighter, the F11F-1 Tiger was the last fighter produced by Grumman Aircraft Corporation until its introduction of the F-14 Tomcat. The Tiger was intended to be a simple, lightweight, air superiority, day interceptor to protect the fleet. Like the F8F Bearcat, it was the smallest airframe possible designed around a given engine. The aircraft was so small that only the tips of the wings folded and folding was accomplished manually.

Thin swept wings incorporating spoilers instead of ailerons coupled with an area-rule (coke-bottle fuselage) design enabled the Tiger to achieve a top speed of over 750 mph and it became the Navy's first operational supersonic fighter. Two variants of the aircraft (F11F-1F) with a more powerful engine and a retractable refueling probe reached a speed of 1386.47 mph in level flight and achieved a world altitude record of 76,828 feet.

First flown in July 1954, the F11F-1's subsequent test trials provided for one of the most bizarre flights when a Grumman pilot managed to shoot himself down with his own bullets. Firing the guns in a dive, the trajectory of the bullets allowed him to overtake them on his pullout causing aircraft damage, an engine flame-out and a crash landing.

Nearly 200 Tigers were produced with some going to the Blue Angel flight demonstration team who retained them for ten years. The balance was assigned to six day-fighter attack squadrons. Having been replaced by the F8U Crusader, the F11Fs saw only brief front-line service and were ultimately sent to Advanced Training Command and Reserve squadrons. *(Information courtesy of the National Museum of Naval Aviation)*.

The Tiger shown above is on display in the PWAM. Another in Blue Angels colors is on display mounted on a pedestal at Walker Field Airport, Grand Junction.

Grumman A-6E Intruder

The Intruder is a two-seat (side-by-side) sub-sonic carrier or shore-based all-weather attack bomber, capable of day and night operations using an internal, fully integrated attack/navigation system. It is powered by a pair of 9,300-lb thrust Pratt & Whitney J52-P-8B turbojets. It has a maximum speed of 644 mph, a service ceiling of 42,400', and a range with maximum payload of 1,011 miles. It is equipped with one under-the-fuselage and four underwing attachment points for weapons and stores and is capable of carrying an external load of 18,000 pounds. *(The Complete Encyclopedia of World Aircraft).*

The A-6E was an advanced upgraded development of the A-6A flown by the US Navy and Marine Corps. The first of the A-6 series (A-6A) were delivered to the Navy in 1963 and to the Marines in 1964. The first operational squadron to receive the A-6As was VA-75 that began its support of US forces in Vietnam flying initially from USS Independence. The Digital Integrated Attack and Navigation Equipment (DIANE) and its subsystems incorporated into the aircraft enabled the crew to attack preselected targets or targets of opportunity at night or under adverse weather conditions without the crew having to look out of the cockpit from launch to recovery.

The Intruder had an excellent slow-flying capability and a combat radius of 1,100 miles. Grumman Aerospace Corporation produced 693 Intruders, which flew for the Navy and the Marines between 1963 and 1967. The Navy's experience in the Korean War showed the need for a new long-range strike aircraft with a high subsonic performance at tree-top height to permit under-the-radar penetration of enemy defenses and capable of finding and hitting small targets in any weather. The result was the Grumman A-6 Intruder.

Manned by a pilot and bombardier/navigator seated side-by-side, the A-6 was designed to be powered by two engines which gave the Intruder far better reliability and performance than a single engine design and was heartily endorsed by flight crews because of the added margin of safety. The concept of incorporating tilting tailpipes to provide for a STOL (Short Takeoff and Landing) was abandoned when it was determined that tilting tailpipes only reduced landing speeds by 7 mph at normal approach weights which the Navy considered acceptable.

Properly used, the Intruders produced disproportionate results as illustrated when two A-6s made a night strike with 26-500-lb bombs against a North Vietnam power plant. The damage was such that the Vietnamese were convinced that B-52s had been at work. A Navy Cross was awarded for this mission as was done for a number of other similar ones. *(Information courtesy of the National Museum of Naval Aviation).*

An Intruder is on display mounted on a pedestal at Walker Field Airport, Grand Junction. It wears the colors of Marine All Weather Attack Squadron VMA-AW-533 during its Vietnam deployment to the "Rose Garden" at Nam Phong, Thailand in 1972.

The last flight of this Intruder (Serial No. 154131), was in April 1994, when it was decommissioned and flown to Walker Field by Navy Squadron VX-5, China Lake Naval Air Weapons Station for use as a static display. It is dedicated to the Marine Aviators of VMA-AW-5433 who were lost in combat over Vietnam in 1972, Captain Len Robertson (Pilot) and Lieutenant Al Kroboth (Bombardier/Navigator); Captain Rob Peacock III (Pilot) and Captain Marshall Price (Bombardier/Navigator); Capt. Jim Chipman (Pilot) and Lieutenant Ron Forrester (Bombardier/Navigator). Lieutenant Kroboth was later recovered after serving as a POW. The names of Captain's Peacock and Price appear under the canopy of the aircraft on display.

Grumman OV-1D Mohawk

The OV-1 Mohawk is a two-seat multi-sensor observation aircraft powered by a pair of 1,400-lb Avco Lycoming T53-L-701 turboprops. It has a maximum speed of 289 mph and a maximum range of 1.011 miles. The OV-1C Mohawk is equipped with an AN/AAS-24 infra-red (IR) surveillance system. The OV-1D Mohawk is equipped with side loading doors able to accept a pallet with side-looking airborne radar (SLAR), IR or other sensors. In addition to newly-built OV-1Ds, many OV-1Bs and OV-1Cs were converted to OV-1D standard. *The Complete Encyclopedia of World Aircraft.*

During the Vietnam War, Army OV-1's flew out of Udorn Royal Thai Air Base and conducted nightly reconnaissance patrols of the Plain of Jars (or Barrel Roll) area of Laos. The IR equipped OV-1C's could detect heat from the engines of trucks and even camp fires. The SLAR equipped OV-1B's were used to detect moving objects. In many cases, the targets identified by the OV-1's were gone by the time the recorded reconnaissance data was gathered, interpreted and analyzed. The idea was to relay the near real-time target data appearing on the OV-1 monitors to AC-119K gunships operating in the area. The Mohawk was also called the Spud. (*Information courtesy of the USAF Museum*).

Grumman F-14A Tomcat
(HAS Photo)

The Tomcat is a two-seat carrier-based multi-role fighter powered by two 20,900-lb thrust Pratt & Whitney TF30-P-412A afterburning turbofans. It has a maximum speed of 1,564 mph, a service ceiling of 68,900' and a combat range of 2,000 miles. It is armed with one General Electric M61A-1 20-mm cannon in the forward fuselage and can carry various combinations of bombs or missiles, including the Phoenix, Sidewinder and Sparrow. It can also carry a tactical reconnaissance pod or ECM equipment. *(The Complete Encyclopedia of World Aircraft).*

Failure of the General Dynamics F-111B to meet US Navy requirements for an advanced carrier-based air superiority fighter left a significant gap in the Navy's inventory. Following its cancellation in April 1968, the Navy launched a new design contest in which the finalists were Grumman and McDonnell with the former declared the winner with its proposed variable geometry, two-seat twin-engined aircraft. Designated the F-14 and eventually named Tomcat, procurement began in 1969 for 700 aircraft for completion in the early nineties. Deliveries to the Navy began in June 1972 with deployment of operational carrier squadrons in 1974.

The ability to sweep its wings aft 43 degrees from the horizontal, coupled with twin 21,000-lb thrust engines enables the F-14 to achieve speeds in excess of twice the speed of sound. The degree of variable sweep is a function of aircraft speed and is computer controlled. As aircraft speed bleeds off for whatever reason (high-g's, landing, etc.), the computer automatically compensates by extending the wing for more lift to prevent a stall from occurring.

The F-14As made a brief appearance over Vietnam, flying protective patrols for helicopters effecting the final evacuation of US forces from Saigon with no opposition from enemy fighters. The Middle East was to become the scene of the Tomcat's combat initiation during encounters with Libyan fighters over the Gulf of Sidra in 1981 when several Sukhoi SU-22 "Fitter" fighters were shot down. In its interceptor roll, the F-14 proved invaluable during the gulf war providing cover for airborne Navy and USAF support aircraft as well as blocking Iraqi aircraft from flying to safe-havens in Iran.

The F-14 with its Phoenix air-to-air missile coupled with airborne early warning aircraft is able to simultaneously intercept, engage and destroy up to five incoming enemy aircraft out to distances in excess of five hundred miles from a carrier task force. The F-14 is now used in the attack role, as well.

Reduction of force requirements and concurrent cuts in defense spending has necessitated gradual replacement by the year 2008 of the F-14s by F/A-18 Hornets. While the latter aircraft lacks the long-range air defense capability of the F-14, it is cheaper to produce and costs less to operate per flight hour. *(Information courtesy of the National Museum of Naval Aviation)*.

The Tomcat shown above is on display inside the WORA&SM.

Hiller OH-23D Raven

The UH-23D is a three-seat military helicopter powered by one 323-hp Avco Lycoming VO-540-A1B flat-six piston engine. It has a maximum speed of 95 mph, a service ceiling of 13,200', and a range of 205 miles. (*The Complete Encyclopedia of World Aircraft*).

In 1949 the Hiller Aircraft Corporation produced a two-seat, two-bladed helicopter that employed a new control system design called Rotor-Matic Control, which made it easy to handle as well as providing a high degree of stability. Designated as the Model 360 (and later by the Navy as the HTE-1 and the UH-12), success came instantly and by 1950 it had become the largest selling helicopter in the world. A model of it became the first helicopter to make a transcontinental commercial flight.

With the beginning of the Korean War, production shifted to the military with the Navy ordering a limited number in comparison to the hundreds purchased by the Army. Equipped with a 200-hp powerplant, it became the most powerful light utility helicopter in the world at that time and was employed by both the Army and Navy as medical evacuation, observation, and utility aircraft in Korea.

About 2,200 production versions of this Hiller helicopter were built in the first production run. In the 1970s it appeared in four versions, two of which employed turbine engines. *(Information courtesy of the National Museum of Naval Aviation)*.

Hawker Sea Fury T Mk.20
(Photo courtesy of Alan Groening)

The Sea Fury is a single-seat carrier-based fighter-bomber powered by one 2,480-hp Bristol Centaurus 18-cylinder radial piston engine. It has a maximum speed of 435 mph, a service ceiling of 34,300', and a range of 680 miles. It is armed with four 20-mm cannon in the wings and can carry eight 60-lb rockets or two bombs in underwing racks. *(The Complete Encyclopedia of World Aircraft)*.

The first prototype was flown on 21 February 1945. 615 were built, including 31 for the Royal Australian Navy and 35 for the Royal Canadian Navy. Sea Furies were used to great effect in the ground attack role in Korea in 1950. *(Information courtesy of the Canadian Forces Archives)*.

Joe Thibodeau's Sea Fury flown out of Denver is a former Iraqi fighter in Royal Canadian Navy colors.

Hunting (Percival) P.84 Jet Provost

The Jet Provost is a two-seat basic trainer powered by one 2,500-lb thrust Bristol Siddeley Viper Mk. 202 turbojet. It has a maximum speed of 440 mph, a service ceiling of 36,700', and a range with tip tanks of 900 miles. It can be armed with to .3-in machine-guns, and can carry a variety of weapons on underwing hardpoints. *(The Complete Encyclopedia of World Aircraft).*

Lockheed RB-37 Ventura
(HAS Photo)

The Ventura is a maritime patrol aircraft powered by a pair of 1,700-hp Pratt & Whitney R-2600-13 radial piston engines. The PV-1 variant had a maximum speed of 322 mph, a service ceiling of 26,300', and a range of 1,360 miles. Its prime purpose was for coastal patrol, and it was armed with two fixed .50 cal. guns mounted in the nose, and another two movable .30 cal. guns mounted below. The Ventura was also equipped with an upper turret armed with a pair of .50 cal. machine-guns as well as Plexiglas depressions in the side to provide lateral fire zones for two more guns. It could carry a bomb load of over one ton.

This version of the Lockheed 37 light bomber was first accepted in September 1941 and was the first bomber to have all metal control surfaces instead of frame and fabric. The Pueblo Air Museum's RB-37 aircraft shown above was transported by truck from the Training Research Center at St. Paul, MN and is on loan from the USAF Museum. *(Information courtesy of the PWAM)*.

Lockheed Model 18/C-60A-5 Lodestar

The Lodestar is a modified civil transport powered by a pair of 875-hp Wright R-1820-87 radial piston engines, it had a maximum speed of 351 km/h, a service ceiling of 20,400', and a range of 2,897 km. *(Information courtesy of the Canadian Forces Archives)*.

The USAAF operated Lodestars during World War II; 102 were taken over from civil airlines as C-56, C-57, C-59 and C-60. Also 324 were purchased new as C-60's. The U. S. Navy also purchased 96 under the designation R-5O (the letter O). The Vega Ventura maritime patrol/light bomber aircraft was based on the Lodestar design. It was designated PV-1 by the Navy and B-34 by the USAAF. *(Information courtesy of the Castle Air Museum)*.

Lockheed P-38L Lightning

The P-38L Lightning was the final Lockheed production version of the twin-tailed single-seat fighter. It was powered with a pair of 1,475-hp Allison V-1710-11/-113 engines. It had a maximum speed of 414 mph, a service ceiling of 44,000', and a range of 450 miles. It was armed with four .50 cal machine-guns and one 20mm cannon mounted in the nose, and could carry up to 3,200-lbs of bombs. *(The Complete Encyclopedia of World Aircraft)*.

The Lightning was designed in 1937 as a high-altitude interceptor. The first one built, the XP-38, made its public debut on February 11, 1939 by flying from California to New York in seven hours. Because of its unorthodox design, the airplane experienced growing pains and it required several years to perfect it for combat. Late in 1942, it went into large-scale operations during the North African campaign where the German Luftwaffe named it Der Gabelschwanz Teufel (The Forked-Tail Devil).

Equipped with droppable fuel tanks under its wings, the P-38 was used extensively as a long-range escort fighter and saw action in practically every major combat area of the world. A very versatile aircraft, the Lightning was also used for dive-bombing, level bombing, ground strafing and photo reconnaissance missions. By the end of production in 1945, 9,923 P-38s had been built. *(Information courtesy of the USAF Museum)*.

On 8 August 1942, First Lieutenant Edward J. Peterson, Operations Officer for the 14th Photo Reconnaissance Squadron and a native of Colorado, crashed while attempting to take off from the airfield when the left engine of his twin engine F4 (a reconnaissance variant of the P-38 Lightning) failed. A base fire department crew rescued Lt. Peterson from the burning wreckage. Unfortunately, Lt. Peterson sustained significant burns and died at a local hospital that afternoon, thereby becoming the first Coloradan killed in a flying accident at the airfield. Consequently, on 13 December 1942, officials changed the name of the Colorado Springs Army Air Base to Peterson Army Air Base in honor of the fallen airman.

Lockheed P-80 Shooting Star

The Lockheed P-80 Shooting Star was one of American's first single-seat jet fighters. The P-80 was powered by an Allison J33-A-35 turbojet, with a maximum speed of 600 mph and a range of 1,345 miles. The first flight made by an XP-80 was made on 9 January 1944, followed by a P-80 equipped with the J33 engine, which was flown 11 June 1944. After June 1946 the Shooting Star was designated F-80. This was the US Air Force's first true jet-powered fighter of the jet air age. *(The Complete Encyclopedia of World Aircraft)*.

Production delivery began in August 1943. On 25 June 1950, the F-80C went to the War in Korea. On 8 November, a pilot of an F-80C tangled with a Russian MiG-15 in the first jet-to-jet dogfight and emerged victorious. By the time production halted on the Shooting Star and its various versions including the TF-80C, there had been 1,714 built. Another version was designated as the T-33 and these were used to train the first jet pilots. There were a total of 5,691 T-33s built. *(Information courtesy of the USAF Museum).*

The PWA Museum's F-80 shown above was restored from the remains of an F-80 that was left on the Aircraft Museum grounds. Dick Wood of Denver did the restoration over a period of 4 years. The markings are from the 8th Fighter-Bomber Wing where the aircraft served for 8 years of its service. *(Information courtesy of the PWAM).*

Lockheed T-33A & B Shooting Star
(HAS Photo)

The two-place T-33 jet was designed for training pilots already qualified to fly propeller-driven aircraft. It was developed from the single-seat F-80 fighter by lengthening the fuselage slightly more than three feet to accommodate a second cockpit. Originally designed the TF-80C, the T-33 made its first flight in March 1948. Production continued until August 1959 with 5,691 T-33s built. In addition to its use as a trainer, the T-33 has been used for such tasks as drone director and target towing, and in some countries even as a combat aircraft. The RT-33A version, reconnaissance aircraft produced primarily for use by foreign countries, had a camera installed in the nose and additional equipment in the rear cockpit.

The T-33 is one of the world's best-known aircraft, having served with the air forces of more than 20 different countries for almost 40 years. Many were supplied to foreign nations under the Military Aid Program, and are still in use throughout the world. A total of nearly 6,000 were built. The Canadian Forces used a license-built version with a Rolls-Royce Nene engine. *(Information courtesy of the USAF Museum).*

The T-33A shown above is on display at the PA&SM in the Air Park. The PWA Museum's T-33B was donated in 1979 by The US Naval Air System Command at Washington D.C. and was transported to Pueblo by truck from The Military Storage and Disposition Center at Davis Monthan AFB, AZ. It wears the colors of the USMC. Another is on display in the WORA&SM. *(Information courtesy of the PWAM).*

Lockheed F-94C Starfire
(HAS Photo)

The Starfire is a two-seat all weather jet fighter powered by one 6,350-lb thrust Pratt & Whitney J48-P-5 turbojet and 8,750-lb thrust with afterburning. It has a maximum speed of 640 mph, a service ceiling of 51,400', and a range of 805 miles. Its armament consisted of 24 folding-fin rockets in the nose, plus 24 similar rockets in two wing pods. *(The Complete Encyclopedia of World Aircraft).*

The F-94 series all-weather interceptors were developed from the Lockheed F-80 Shooting Star. The prototype F-94 first flew on July 1, 1949. The Starfire was subsequently produced in the -A, -B, and -C series. The F-94C (originally designated the F-97A) was a fundamental redesign of the F-94B and made its first flight on the 18th of January 1950. Improvements in the F-94C included a higher thrust engine, single point refueling, a redesigned wing, a swept-back horizontal stabilizer, upgraded fire-control and navigation systems and, later, mid-wing rocket pods. Twenty-four rockets were carried in the nose in a ring around the radome, shielded by retractable doors, with an additional 24 in the wing pods, if installed. The F-94C carried no guns.

Starfires were employed in the air defense of the Continental U.S. in the 1950s. In the F-94A form, they served as the first all-jet all-weather interceptor for the Air Defense Command. The last F-94Cs were withdrawn from USAF service in 1959. *(Information courtesy of the USAF Museum).*

The Starfire shown above is on display at the PA&SM Air Park.

Lockheed F-104C Starfighter

The Starfighter is a single-seat multi-mission fighter powered by one 15,600-lb afterburning thrust General Electric J79-GE-11A turbojet. It had a maximum speed of 1,146 mph, a service ceiling of 50,000', and a range of 1.081 miles. *(The Complete Encyclopedia of World Aircraft).*

Designed as a supersonic superiority fighter, the F-104 was produced in two major versions. Armed with a six-barrel M-61 20mm Vulcan cannon it served as a tactical fighter and, equipped additionally with heat-seeking Sidewinder missiles, as a day-night interceptor. Development of the F-104 began in 1952 and the first XF-104 made its initial flight in 1954. The F-104A was the first production version initially fitted with J79-GE-3A engines, it was later retrofitted with J79-GE-3B turbojets, a ventral fin and blown flaps, with 153 built. On May 18, 1958, an F-104A set a world speed record of 1,404.19 mph. *(Information courtesy of the USAF Museum)*.

The Starfighter was the world's first operational fighter aircraft to exceed Mach 2. The aircraft was built to wrap around a powerful turbojet engine. Its elongated fuselage and short-span wings made it look more like a missile than an aircraft. It does, however, need a long take-off run. Designed as a supersonic superiority fighter, the F-104 was produced in two major versions. Armed with a six-barrel M-61 20mm Vulcan cannon it served as a tactical fighter and, equipped additionally with heat-seeking Sidewinder missiles, as a day-night interceptor.

Development of the F-104 began in 1952 and the first XF-104 made its initial flight in 1954. On the 18th of May 1958, an F-104A set a world speed record of 1,404.19 mph, and on the 14th of December 1959, an F-104C set a world altitude record of 103,395 feet. The Starfighter was the first aircraft to hold simultaneous official world records for speed, altitude and time-to-climb. The USAF procured about 300 Starfighters in one and two-seat versions. In addition, more than 1,700 F-104s were built in the U.S. and abroad under the military aid program for various nations including Canada, West Germany, Italy, Norway, the Netherlands, Belgium, Denmark, Greece, Turkey, Spain, Taiwan and Japan. *(Information courtesy of the USAF Museum)*.

The F-104C was a tactical strike fighter version built for the USAF Tactical Air Command. It was powered by one 15,800-lb afterburning thrust J79-GE-7 engine and was armed with Sidewinder missiles or with bombs or rocket pods for conventional or nuclear strike missions. It was fitted with an inflight-refueling probe. 77 were built.

On 14 December 1959, an F-104C set a world altitude record of 103,395 feet. The Starfighter was the first aircraft to hold simultaneous official world records for speed, altitude and time-to-climb. *(The Complete Encyclopedia of World Aircraft).*

There are Starfighters on display at Peterson AFB; the USAF Academy; and in the WORA&SM.

Lockheed 1049 Constellation

The Constellation is a long-range transport aircraft powered by four 3,400-hp Wright 988TC-18EA-02 Turbo-Compound radial piston engines. It has a maximum speed of 377 mph, a service ceiling of 23,700', and a range of 4,940 miles. *(The Complete Encyclopedia of World Aircraft).*

The Constellation is a superb piston-engined airliner, which was first introduced on transatlantic service flights in 1953. The Connie was typical of the long-range piston-engined transports that preceded the jet-age. The fuselage was pressurized to accommodate electronic technicians and their associated mission equipment at high altitudes. *(Information courtesy of The Air Museum Planes of Fame, Chino).*

Lockheed EC-121T Warning Star
(HAS Photo)

The Lockheed EC-121 Warning Star is a variant of the Lockheed Super Constellation C-121 long-range transport aircraft. Utilized for airborne early warning, it employs a radome height-finding radar antenna (called the camel back) on top of the fuselage and a large radome under the fuselage for a 20' search radar. Four 3,400-hp Wright 988TC-18EA-02 Turbo-Compound radial piston engines power the Warning Star. It had a maximum speed of 377 mph (without radome), a service ceiling of 23,700', and a range of 4,940 miles. The Warning Star is the forerunner of the E3A AWACs. *(The Complete Encyclopedia of World Aircraft)*.

Of the 142 C-121s ordered by the Navy, several in the transport version fulfilled the Navy's obligation to the Military Air Transport Service while the balance participated with the Air Force flying Early Warning Barriers over the Atlantic and Pacific Oceans. A portion of this role ended in 1965, when an advanced radar system assumed the responsibility over the northern approaches to the North Atlantic—a mission previously assigned to Navy EC-121 squadrons for ten consecutive years. This Jack-of-all-Trades aircraft was also employed as both a weather, electronic- and photo-reconnaissance aircraft. During Vietnam both the Air Force and Navy employed the EC-121 in a variety of roles ranging from aircraft command and control and early warning to electronic surveillance. Called Connies by the pilots who flew them, a replacement for them in these roles was never found.

The Warning Star also operated in a more peaceful mission in the form of a Navy C-121J nicknamed the Blue Eagle that served as an airborne radio and television transmitter for the armed forces network in South Vietnam. While this particular model served as a transmitter of news, another Warning Star became a maker of news on 14 April 1969 when a Navy EC-121 was shot down over the Sea of Japan by North Korean aircraft with the loss of the entire thirty-one man crew. This attack prompted the United States to deploy Task Force 71 to Asian waters in protection of future flights. *(Information courtesy of the National Museum of Naval Aviation)*.

The EC-121T Warning Star (Serial No. 52-3425), shown above is on display at the PA&SM Air Park. This aircraft was flown to Peterson AFB in the fall of 1978 after flying Icelandic missions as late as May 1978. It has an extensive history available for review in the museum. The interior of the aircraft is virtually intact from its last mission, and the museum's restoration plans include eventually opening the interior for visitation after restoration is complete. *(Information courtesy of the PA&SM)*.

Other Warning Stars have been preserved at Helena, MT (Serial No. 52-3417); at the Combat Air Museum in Topeka, KS (Serial No. No. 52-3418); and at Camarillo, CA (Serial No. 53-0548). One Warning Star is still flying with the Pima Air Museum, AZ (Serial No. 53-0554).

Lockheed P2V-5 Neptune
(HAS Photo)

The Neptune is a long-range maritime patrol aircraft powered by a pair of 2,800 hp Wright R-3350-24W engines. It had a maximum speed of 303 mph, a service ceiling of 22,000', and a range of 3,685 miles. *(The Complete Encyclopedia of World Aircraft)*.

After close contact with Navy patrol squadrons, Lockheed concluded that a new, shore-based, ASW long-range airplane with a greater ordnance load capacity was needed, and began designing a prototype in 1941 as a private venture. World War II delayed the first flight of a prototype until 1945 following an order from the Navy for two of them. Delivery of production models began in 1947. Designated the P2V Neptune, it grossed out at 61,000-lbs, and was armed with three .50 caliber machine gun turrets (later converted to 20-mm cannons), an enclosed bomb-bay capable of carrying two torpedoes or twelve depth charges, and had provisions for sixteen underwing rockets.

The P2V-5 was the most numerous Neptune variant produced and would serve as the basis for the largest number of sub-type modifications developed of any of the Neptune series. First flown on the 29th of December 1950, twenty-three Model 426-42-06s were built. The new P2V-5 replaced the six cannon solid nose cone of the P2V-4 with an Emerson Aero 9B turret armed with a pair of 20mm cannon. The 20mm tail turret as well as the .50 caliber machine gun armed dorsal turret remained unchanged from the P2V-4. The P2V-5 was capable of carrying an 8,000lb load of mines, torpedoes, bombs or depth charges and sixteen rockets on under-wing launch stubs, later reduced to 8 with the addition of the J-34s.

An APS-20 radar was mounted in the underbelly position just behind the nose wheel doors. The wing tip fuel tanks were enlarged and were moved from an under wing tip position on the P2V-4, to the outside center of the wing tips and could carry an additional 350 gallons of fuel in each tank. In an emergency, the new tip tanks could be jettisoned. In addition, the forward portion of the starboard wingtip tank housed a powerful searchlight with a moveable reflector and arc element that was linked by servomotors to the nose turret guns. The port wingtip tank had APS-8 search radar mounted in its forward portion. The increase in ASW/ECM equipment increased the aircrew to nine men. The Wright Cyclone R-3350-30W engines remained unchanged from the P2V-4. The weight of the additional equipment without an increase in power caused the top speed to drop to 341 mph. However, the added fuel capacity increased the P2V-5's range to 4,750 miles.

The Neptune was designed in 1942 by Vega Aircraft Corp., a subsidiary of Lockheed, as a land based photo-reconnaissance and anti-submarine duty aircraft. Lockheed dissolved Vega in 1943 and an XP2-1 made its first flight in May 1945. A P2V named the "Truculent Turtle" flew from Perth Australia Sept. 29, 1946 to Columbus Ohio, 11255.6 miles, in 55 hours and 17 minutes breaking the world distance record without refueling. Although designed for land operation this aircraft was flown from carriers. Successive models progressed from the P2V-1 to the P2V-7. In 1962 the P2-Vs were re-designated as P-2s. The P2V-s became the P-2 E and the P2V-7 became the P-2H. More than 1,000 Neptunes were built for the USN and other services for maritime patrol duties. The land-based Neptune led to the development of the current US P-3 Orion. *(Information courtesy of the National Museum of Naval Aviation).*

The PWAM's aircraft shown above was at one time equipped with jet engines attached outboard of the regular engines. They were removed prior to being donated by the Navy Air Reserve Detachment of Buckley Field, CO. *(Information courtesy of the PWAM).*

Martin EB-57E Canberra

(HAS Photo)

The Canberra is a twin-engine night intruder powered by a pair of 7,200-lb thrust Wright J65-W5 turbojets. It has a maximum speed of 582 mph, a service ceiling of 48,000', and a range of 2,300 miles. The Canberra is armed with eight .50-cal machine-guns or four 20-mm cannon, 16 underwing rockets and up to 6,000-lbs of bombs carried in an internal bomb-bay. *(The Complete Encyclopedia of World Aircraft)*.

The English Electric Canberra was the first jet bomber to be built in Britain and the first to serve with the RAF. The USAF made use of the Canberras as bomber and reconnaissance aircraft over Southeast Asia. First flown on 23 April 1950, the Canberra held the World Height Record of 65,889 feet in August 1955. The B-57 is a modified version of the English Electric Canberra, which was first flown in Britain on the 13th of May 1949, and later produced for the Royal Air Force (RAF). After the Korean Conflict began in 1950, the USAF looked for a jet medium bomber to replace the aging Douglas A-26 Invader. In March 1951, the USAF contracted with the Glenn L. Martin Co. to build the Canberra in the U.S. under a licensing agreement with English Electric. The Martin-built B-57 made its first flight on the 20th of July 1953, and when production ended in 1959, a total of 403 Canberras had been produced for the USAF. *(Information courtesy of the USAF Museum)*.

The Canberra shown above is on display in the PA&SM Air Park, another may be viewed in the WORA&SM.

Martin Marietta SV-5J (X-24A) Lifting Body
(HAS Photo)

A lifting body is an aerodynamic vehicle, which derives its lift from the shape of its body instead of from conventional wings. Martin Marietta first explored the concept by testing three of these small unmanned vehicles. The program was successful and they were soon followed by the manned X-24A, which was powered by a Thiokol XLR-11 turbo-rocket engine. The lifting body shown above has been on display outside the Aeronautics laboratory at the USAF Academy since 9 April 1980.

McDonnell Douglas A-4D-2 Skyhawk

The Skyhawk is a single-seat carrier-based attack bomber, powered by one 7,800-lb thrust Pratt & Whitney J65-W-16A turbojet. The A-4D-2 is an improved variant of the Skyhawk powered by one 8,500-lb thrust Pratt & Whitney J52-P-6A turbojet and modified with two additional underwing hardpoints to allow a maximum weapon load of 8,300-lbs. *(The Complete Encyclopedia of World Aircraft).*

Ed Heinemann designed the Skyhawk in response to a Navy requirement for a fast (but compact) long-range, light-weight carrier jet aircraft capable of delivering a nuclear weapon. Prototype test results in 1954 confirmed that the Skyhawk exceeded all of the Navy's criteria. Because of its small size (wing span less than that of a Piper Cub) and ease with which flight deck personnel could handle it in comparison with other jet aircraft, A-4Ds became known variously as either Scooters, Tinker Toys or Heinemann's Hot Rod.

An A-4 set a world speed record of over 695-mph in 1959 for class C aircraft over a 500 km course. Fitted with two 150 gallon under wing drop tanks, two A-4Ds flew 2,082 miles non-stop without in-flight refueling in a demonstration of its long-range capability. While original specifications limited the aircraft to 30,000-lbs fully loaded, various weight-saving measures reduced that to 25,000-lbs. This was accomplished by elimination of a heavy duty battery in favor of a fuse-lage stored wind driven generator; back-up hydraulic system elimi-nated by gravity dropped gear; and installation of a simplified air conditioning system one third the weight of those then available.

The A-4 was stress limited to 24,500-lb total weight for catapult launches, and 5,000-lb ordnance loads on a center line and four wing racks ranging from conventional bombs, to sophisticated weapons such as the Gatling gun, Bullpup, Walleye, Shrike and, in one case, Sidewinder air-to-air missiles. Built into the aircraft were two 20mm cannons. The Skyhawk participated in the first raids of the Vietnam War and became one of the primary strike aircraft thereafter until replaced by the A-7 Corsair in the 1969 time-frame. An A-4C is cred-ited with shooting down a MiG-17 over Vietnam. The A4Ds endured the most losses of any carrier-based aircraft in Vietnam with the loss of 195 of them in combat including those piloted by Senator John McCain and Vice Admiral James Stockdale as well as the first two vic-tims of surface-to-air guided missiles during that conflict. The A-4 also saw considerable combat action during the Arab/Israel and Falkland Island wars.

Nearly 3,000 A-4s were produced from 1956 to 1979 for use by the Navy and Marine Corps as well as Australia, Israel, Argentina and Kuwait. A two-seat trainer version was still used by the Navy until late 1999. *(Information courtesy of the National Museum of Naval Aviation).*

The Pueblo Air Museum Skyhawk was donated in 1979 by the US Naval Air Systems Command at Washington D.C. and was transferred by truck from Davis Monthan AFB, AZ. It is now suspended from the ceiling of the museum's new aircraft hangar. *(Information courtesy of the PWAM).*

McDonnell F-101A Voodoo
(HAS Photo)

The F-101A Voodoo is a single-seat all-weather long-range interceptor powered by two 14,880-lb thrust afterburning Pratt & Whitney J57-P-55 turbojets. It had a maximum speed of 1,221 mph, a service ceiling of 40,000' and a range of 1,550 miles. It was armed with two MB-1 Genie missiles with nuclear warhead and four AIM-4C, -4D or 04G Falcon missiles or six Falcon missiles. *(The Complete Encyclopedia of World Aircraft).*

The Pueblo Air Museum's F-101A shown above made its first flight May 1953 and was assigned to the USAF's 81st tactical wing as a low-level fighter/bomber. Due to its range (without refueling) of 1,700 miles it was the first long-range escort fighter. The Museum's aircraft is the first F-101 manufactured and is the only F-101 owned by a civilian. This aircraft is on loan to the City of Pueblo by its owner Dennis Kelsey. This aircraft is undergoing restoration. As time permits, volunteer workers will restore it to at least aesthetic standards. *(Information courtesy of the PWAM)*.

McDonnell F-101B Voodoo

The F-101A Voodoo is a single-seat and the F-101B Voodoo is a two-seat all-weather long-range interceptor powered by two 14,880-lb thrust afterburning Pratt & Whitney J57-P-55 turbojets. It had a maximum speed of 1,221 mph, a service ceiling of 40,000' and a range of 1,550 miles. It was armed with two MB-1 Genie missiles with nuclear warhead and four AIM-4C, -4D or 04G Falcon missiles or six Falcon missiles. *(The Complete Encyclopedia of World Aircraft)*.

Developed from the XF-88 penetration fighter, the F-101 originally was designed as a long-range bomber escort for the Strategic Air Command (SAC) (now known as Strategic Command). However, when high-speed, high-altitude jet bombers such as the B-52 entered active service, escort fighters were not needed. Therefore, before production began, the F-101's design was changed to fill both tactical and air defense roles. The F-101 made its first flight on Sep. 29, 1954. The first production F-101A became operational in May 1957, followed by the F-101C in September 1957 and the F-101B in January 1959. By the time F-101 production ended in March 1961, McDonnell had built 785 Voodoos including 480 F-101Bs, the two-seat, all-weather interceptor used by the Air Defense Command. In the reconnaissance versions, the Voodoo was the world's first supersonic photo-reconnaissance aircraft. These RF-101s were used widely for low-altitude photo coverage of missile sites during the 1962 Cuban Missile Crisis and during the late 1960s in Southeast Asia.

The F-101 lineage included several versions: low-altitude fighter-bomber, photo-reconnaissance, two-seat interceptor and transition trainer. To accelerate production, no prototypes were built, the first Voodoo, an F-101A, made its initial flight on 29 September 1954. When production ended in March 1961, nearly 800 Voodoos had been built. Development of the unarmed RF-101, the world's first supersonic photo-reconnaissance aircraft, began in 1956 while 35 RF-101As and 166 RF-101Cs were produced, many earlier single-seat Voodoos were converted to the reconnaissance configuration. (*Information courtesy of the USAF Museum*).

USAF Voodoos are on display in the PA&SM Air Park and in the WORA&SM.

McDonnell CF-101B Voodoo

The RCAF operated 56 two-seat long-range all-weather interceptor Voodoos as part of NORAD from July 1961. They also served with the Canadian Forces under the designation CF-101B. They were also provided with ten TF-101Bs redesignated CF-101F after being upgraded with an infra-red detection and fire control system. Ten years later, the surviving Canadian Voodoos were exchanged for a number of refurbished aircraft, and four CF squadrons continued to operate them until they were replaced with McDonnell Douglas CF-18s. (*Information courtesy of the CF Archives*).

A Canadian Voodoo is on display in the PA&SM Air Park. This aircraft was retired from No. 414 Squadron, North Bay Ontario. It was flown to Peterson AFB in the summer of 1984. It was officially handed over on 25 October 1985, and refurbished on 3 June 1988. (*Information courtesy of the PA&SM*).

McDonnell-Douglas F-4C Phantom II
(HAS Photo)

The two-place F-4C Phantom II is an all-weather, multi-role fighter with advanced radar and missile armament. It is powered by a pair of 17,000-lb afterburning thrust General Electric J79-GE-15 turbojets, with a maximum speed of 1,485 mph, a service ceiling of 62,250', and a combat radius of 595 miles. It also incorporates numerous system changes from the earlier versions. *(The Complete Encyclopedia of World Aircraft).*

The Phantom II was originally developed for U.S. Navy fleet defense and entered service in 1961. The USAF evaluated it for close air support, interdiction, and counter-air operations and, in 1962, approved a USAF version. The USAF's Phantom II, designated F-4C, made its first flight on 27 May 1963. Production deliveries began in November 1963, and 635 were built.

The armament loaded on a typical F-4C consists of four AIM-7E and four AIM-9B air-to-air missiles, and eight 750 lb. Mk. 117 bombs. The aircraft could also carry two external 370-gallon fuel tanks on the outboard pylons and one ALQ-87 electronic countermeasures (ECM) pod on the right inboard pylon. This was one of the typical armament configurations for the F-4C during the Vietnam War in the summer of 1967.

The Phantom is one of the most versatile aircraft in USAF and USN history, serving in fighter-attack, reconnaissance and interceptor roles. Based on its combat record the Phantom may be considered to be the most significant American fighter to take to the skies in combat since WWII. The USAF and the USN Phantoms scored 72% of all victories over North Vietnam between 1965 and 1973.

The Phantom is the only aircraft credited with six MiG kills since the Korean War. Captain Steve Ritchie, made his first and fifth kills a Phantom which is currently on display at the USAF Academy in Colorado Springs. *(Information courtesy of the USAF Museum).*

The Phantom II shown above is on display at Fremont County Airport.

McDonnell-Douglas F-4E Phantom II

The F-4E was a major production version of the Phantom, introducing a pair of more powerful J79-GE-17 turbojets, increased fuel capacity, a redesigned nose with smaller APQ-120 radar, leading edge slats to improve maneuverability, and a 20-mm multi-barrelled cannon. 1,405 were built. *(Complete Encyclopedia of World Aircraft).*

An F-4E on display in the WORA&SM.

McDonnell Douglas F-15A Eagle
(HAS Photo)

The Eagle is a single-seat all-weather air-superiority fighter powered by two Pratt and Whitney F-100 turbofan engines which produce high performance, high maneuverability, and enable it to quickly achieve Mach 2.5. Each engine has 23,000 pounds of thrust. The Eagle can climb to 50,000 feet in one minute. It is armed with one M61A1 20mm six-barreled cannon and four AIM-9 Sidewinder, four AIM-7 Sparrow or eight AMRAAM air-to air missiles. The Eagle can also carry 6,000-lbs of weapons externally. It is an extremely maneuverable aircraft, which entered UASF service on 14 November 1974. *(The Complete Encyclopedia of World Aircraft).*

First flown on 27 July 1972, the Eagle began entering the USAF inventory on 14 November 1974. It was the first U.S. fighter to have engine thrust greater than the normal weight of the aircraft, allowing it to accelerate while in a vertical climb. This, combined with low aircraft weight compared to wing area, made the Eagle highly maneuverable. The Eagle has been produced in single-seat and two-seat versions. During Operation Desert Storm F-15Cs conducted counter-air operations over Iraq. They escorted strike aircraft over long distances and scored 36 of the 39 USAF aerial victories during the conflict. The F-15C was also used to search out and attack Scud ballistic missile launchers. *(Information courtesy of the United States Air Force Museum).*

The F-15A shown above is on display at the USAF Academy and is maintained by the "4ᵗʰ Cadet Group." Another F-15 is on display at the PA&SM Air Park.

McDonnell Douglas F-15B/C/D/E Eagle

This year the United States Air Force celebrated its Fiftieth anniversary. For twenty-five years, half of the time that the Air Force has been in existence, one aircraft has been at the forefront of tactical fighter aircraft, the McDonnell Douglas F-15 *Eagle*. The *Eagle's* history is long and distinguished. It began as an Air Force fighter study in the early 1960s and was known as the Fighter Experimental (FX). By 1967 the Air Force began development of a new high performance fighter aircraft that would be extremely agile and would be capable of gaining and maintaining air superiority through air-to-air combat. The new design had to be optimized for combat with the power and agility to overcome any current or projected Soviet threat. The F-15 was the first air-to-air fighter requested by the Air Force since the F-86 Sabre. The resulting F-15 *Eagle* had an unequalled combination of performance, firepower, and avionics. It was the benchmark—the plane to beat.

To succeed in the air-to-air role, a plane needs the right airframe in combination with strong powerplant and avionics. The plane's designers understood this and stretched technology to the limits. It was determined that a very low wing loading combined with heavy thrust from the engines would be required. US fighter aircraft of the period were going faster (Mach 2 plus), but were heavy and lacked maneuverability compared to their Soviet counterparts. When combined with a capable airframe, better maneuverability can be achieved by maximizing thrust, thereby maximizing energy. The Pratt & Whitney F100 Turbofan engine provides the needed thrust. Each engine is capable of producing 15,000 pounds of thrust at maximum power, and 25,000 pounds of thrust in afterburner. This gives the *Eagle* a total of 50,000 pounds of thrust. In other words, a nominally loaded F-15 *Eagle* of 48,000 pounds has a thrust-to-weight ratio of 1.04 pounds of thrust to each pound of aircraft weight. Thrust of this caliber allows an F-15 to accelerate while going straight up! A specially modified F-15A *Eagle* known as the "*Streak Eagle*" was able to outclimb a Saturn V Moon Rocket to almost 60,000 feet. This same aircraft flew to 98,430 feet (30,000 meters) in 207.80 seconds (less than 3 minutes and 30 seconds).

The lightly loaded airframe is combined with an equally impressive flight control system. A hydraulically actuated, mechanically controlled flight control system is augmented by an electronic system known as the Control Augmentation System (CAS). This system takes the stick inputs from the pilot and deflects the flight controls in the proper direction at the proper rate for optimal aircraft handling. This system allows the pilot to fly the aircraft to the limits of its capabilities without losing control of the aircraft. The CAS can also actuate the flight controls via pilot input if the hydro-mechanical system is damaged.

In order to win air-to-air battles, the pilot must be able to see, shoot, evade, and destroy the adversary first. The *Eagle* has an impressive array of weapons and avionics, which allow it to get the advantage. The APG-63 and 70 radars allow crews to see targets that are as far away as 100 miles. These "Eyes" are able to ferret out the targets even if the targets are flying at high speeds at low altitudes. A Tactical Electronic Warfare System (TEWS) lets the aircrew know if any threat is present. The Heads-up-Display (HUD), and the Hands on Throttle and Stick (HOTAS), allow the Pilot to select, track and shoot the adversary without having to look back into the cockpit.

The impressive avionics suite is backed up by an equally impressive weapons capability. For close air-to-air combat the *Eagle* carries the six barrelled 20mm M61A1 Vulcan cannon. The Vulcan fires rounds at rates of 66 or 100 rounds per second. Further distances are covered by the heat seeking AIM-9 Sidewinder, AIM-7 Sparrow, and the deadly AIM-120 Advanced Medium Range Air-to-Air Missile.

The F-15 *Eagle* has been produced in five models. The single seat A and C models, the two seat B and D models, and the formidable F-15E *Strike Eagle*. The A through D models are air-to-air versions but the E Model can carry out an air-to-ground mission in addition to the original air-to-air mission.

The F-15E is capable of delivering over 20,000 pounds of air-to-ground ordinance while travelling at high speeds at very low altitudes (as low as 100 feet) at night. To perform this role the aircraft utilizes the sophisticated LANTIRN system (Low Altitude Navigation Targeting Infrared for Night). This multi-role aircraft was conceived early in the development of the F-15 because it was easier to convert a thoroughbred air-to-air fighter into a ground attack platform than to complete this function in the opposite manner. The rear cockpit has become a dedicated air-to-ground crew station, housing several sophisticated multi-function display screens and two hand controllers. Added fuel is carried in contoured side fuselage tanks that increase the range without adding excessive aircraft drag. These conformal fuel tanks, fully integrated into the aircraft structure, contain hard point stations for additional air-to-ground ordinance loads.

The F-15 continues to serve with distinction with not only the U.S. Air Force but in Israel, Saudi Arabia, and Japan as well. In service for a quarter of a century, the F-15 *Eagle*'s combat record includes the Bekah Valley, the Persian Gulf, and Iraq. The *Eagle* also delivered heavy ordinance in Operation Deliberate Force, effectively bringing the Serbian forces back to the negotiating tables. The F-15 served in Operation Deny Flight in Bosnia, and continues to fly over Iraq. This impressive aircraft will continue to soldier on into the 21st century as the Air Force's principal fighter. The F-15 Eagle will continue to be the benchmark, "the plane to beat" until it hands the reigns of air superiority to its successor, the Lockheed F-22 Raptor. *(Information courtesy of Edwards AFB Public Affairs Office).*

Mikoyan-Gurevich MiG-17F/JJ-5
(Photo by John Haslo)

Mikoyan Gurevich MiG-17F/JJ-5 is a two-seat version of the MiG-17 built in China in 1977. It is armed with one 23-mm cannon. Colonel Jack E. Wilhite of Red Storm Airshows, flies one of these aircraft out of Jefferson County airport. The MiG-17 this aircraft is modeled on, is a single-seat day fighter powered by one Klimov VK-1F turbojet with afterburning thrust of 7,452 lb. It had a maximum speed of 711 mph, a service ceiling of 54,460', and a range of 1,230 miles. It is armed with one N-37 37mm and two or three 23mm NR-23 cannon and can carry up to 1,102-lbs of weapons on underwing hardpoints. *(The Complete Encyclopedia of World Aircraft).*

The MiG-17 is a refined version of the famous MiG-15 of the Korean War. Although similar in appearance to the MiG-15, the MiG-17 has more sharply swept wings, afterburner, better speed and handling characteristics and is about three feet longer. The first flight of a MiG-17 prototype took place in January 1950 and production began in late 1951. The first operational MiG-17s appeared in 1952 but were not available in sufficient quantities to take part in the Korean War.

Five versions of the aircraft eventually were produced. The MiG-17 has served in the air arms of at least 20 nations throughout the world, including nations friendly to the US, and was flown against U.S. aircraft during the Vietnam War. Between 10 July 1965, and 14 February 1968, USAF F-105s and F-4s downed 61 MiG-17s. *(Information courtesy of the USAF Museum).*

Nord 3202

The Nord 3202 was a French Army Air Force two-seat observation aircraft that was also used for the evacuation of casualties. It is a single-engine aircraft powered by a 260-hp Potez 4D34 inline piston engine. It has a top speed of 146 mph, and a range of 621 miles. The Nord 3202 provided tandem two-seat enclosed accommodation for the instructor (who sat in the rear seat) and student. One is on display in the WORA&SM. *(Information courtesy of the WORA&SM).*

North American AT-6C/D/F/G Texan

(HAS Photo)

The Texan was a two-seat advanced trainer powered by a 550-hp Pratt & Whitney R-1340-AN-1 radial piston engine. It had a maximum speed of 205 mph, a service ceiling of 21,500', and a range of 750 miles. *(The Complete Encyclopedia of World Aircraft).*

The AT-6 advanced trainer was one of the most widely used aircraft in history. Evolving from the BC-1 basic combat trainer ordered in 1937, 15,495 Texans were built between 1938 and 1945. The USAAF procured 10,057 AT-6s; others went to the Navy as SNJs and to more than 30 Allied nations. Most AAF fighter pilots trained in AT-6s prior to graduation from flying school. Many of the Spitfire and Hurricane pilots in the Battle of Britain trained in Canada in Harvards, the British version of the AT-6. To comply with neutrality laws, U.S. built Harvards were flown north to the border and were pushed across. In 1948, Texans still in USAF service were re-designated as T-6s when the AT, BT, and PT aircraft designations were abandoned. To meet an urgent need for close air support of ground forces in the Korean Conflict, T-6s flew mosquito missions spotting enemy troops and guns and marking them with smoke rockets for attack by fighter-bombers. *(Information courtesy of the USAF Museum).*

The AT-6 Texan was also UK and built for Canada where it was called the Harvard. It was used extensively in the Commonwealth Air Training Plan. *(Information courtesy of the Castle Air Museum).*

The Texan shown above is on display at Buckley AFB.

North American P-51D Mustang
(Photo courtesy of Greg Gedny)

The Mustang is a single-seat interceptor and long-range day-fighter powered by one 1,695-hp Packard Merlin V-1650-7 V-12 piston engine. The F-51D model has a maximum speed of 437 mph, a service ceiling of 41,900', and a range of 2,080 miles. It is armed with six .50 cal machine-guns mounted in the wings, and can carry two 1,000-lb bombs or six .5-in rocket projectiles. *(The Complete Encyclopedia of World Aircraft)*.

The P-51 was designed as the NA-73 in 1940 at Britain's request. The design showed promise and AAF purchases of Allison-powered Mustangs began in 1941 primarily for photo-reconnaissance and ground support use due to its limited high-altitude performance. But in 1942, tests of P-51s using the British Rolls-Royce Merlin engine revealed much improved speed and service ceiling, and in December 1943, Merlin-powered P-51Bs first entered combat over Europe. Providing high-altitude escort to B-17s and B-24s, they scored heavily over German interceptors and by war's end, P-51s had destroyed 4,950 enemy aircraft in the air, more than any other fighter in Europe, while losing 2,500 of their own.

Mustangs served in nearly every combat zone, including the Pacific where they escorted B-29s to Japan from Iwo Jima. Between 1941-5, the AAF ordered 14,855 Mustangs (including A-36A dive-bomber and F-6 photo-reconnaissance versions), of which 7,956 were P-51Ds. During the Korean War, P-51Ds were used primarily for close support of ground forces until withdrawn from combat in 1953. The TF-51D is a two-seat version of the P-51 Mustang. *(Information courtesy of the USAF Museum)*.

The incredible Mustang photo above was taken by Greg Gedny. This Mustang is owned and flown by Joe Thidodeau, and is also shown on the cover.

North American F-86F Sabre

The F-86 Sabre is a single-seat all-weather/night fighter-bomber/interceptor powered by one 7,500-lb thrust afterburning GE J-47-13 turbojet engine. It had a maximum speed of 707 mph, a service ceiling of 54,600', and a range of 835 miles. It was armed with six .50 cal nose-mounted machine-guns. *(The Complete Encyclopedia of World Aircraft)*.

The F-86F had a modified wing and 1,539 were built. It was followed by a redesign of the aircraft as an all-weather/night fighter with the designation F-86D. The Sabre was the first US fighter to be developed using captured German swept-wing technology. It had been discovered that the build-up of shock waves on the leading-edge of the wing could be postponed by sweeping it aftwards, so that the angle formed by the wing's leading edge and fuselage was less than 90 degrees.

The Sabre was the primary opponent of the formidable MiG-15 during the Korean War. Many Sabre variants were built, ranging from the standard day-fighter to the heavily modified Sabre Dog. The F-86D (known briefly as the YF-95A) made its first flight on December

22, 1949. It was developed as an all-weather interceptor version of the famed F-86A, the airplane that won supremacy of the skies from the MiG 15 during the Korean War. *(Information courtesy of the USAF Museum).*

An F-86F was on display at the USAF Academy. It was traded for the A-10 now displayed at Thunderbird Air Park, and has been returned to the USAF Museum.

North American F-86H Sabre

The F-86H is a fighter-bomber version of the Sabre powered by a J73 engine and armed with four .50-cal machine-guns. One is on display in the WORA&SM.

North American F-86L Sabre Jet
(HAS Photo)

The F-86L Sabre Dog, also known as the Sabre Jet, varied from its predecessors by having a large radar for night-fighting within its black nose-cone. The Sabre Dog was also powered by a General Electric J47-GE-17B or -33 turbojet. Sabre Dog L variant was a rebuilt variation of the D model with an increased wingspan and updated avionics. *(The Complete Encyclopedia of World Aircraft).*

The jet's air intake was shifted and became a chin-scoop under the nose. The Sabre Dog was an all-weather day and night fighter. Sabres wore very little paint, as it tended to cut the maximum air-speed by as much as 20 miles-per-hour. The aircraft was equipped with a weapons pod under the nose fitted to fire a salvo of Mighty Mouse rockets. *(Information courtesy of the USAF Museum).*

The F-86L is actually an upgraded F-86D with a longer span wing with extended and slatted leading edges and updated avionics. A total of 981 F-86D aircraft were modified to the F-86L standard. *(Information courtesy of the Pima Air & Space Museum).*

The F-86L Sabre Jet shown above is on display in the PA&SM Air Park. It was manufactured by North American Aviation in California and delivered to the USAF on 12 March 1955 as an F-86D. It served in a great number of fighter units California, Arizona, Washington, Utah, Florida, and Idaho, before being placed in storage at Davis Monthan AFB in Arizona in 1965. *(Information courtesy of the PA&SM).*

North American F-100A Super Sabre

The Super Sabre is a single-seat fighter-bomber powered by one 17,000-lb afterburning thrust Pratt & Whitney J57-P-21A turbojet. It has a maximum speed of 864 mph, a service ceiling of 46,000', and a range of 600 miles. It is armed with four 20mm cannon and can carry a weapon-load of 7,500-lbs on underwing pylons including bombs, missiles and rockets. The F-100A was the first production version of the aircraft. *(The Complete Encyclopedia of World Aircraft).*

The Super Sabre was the USAF's first operational aircraft capable of flying faster than the speed of sound (760 mph) in level flight. It made its initial flight on the 25th of May 1953 and the first production aircraft was completed in October 1953. North American built 2,294 F-100s before production ended in 1959. Designed originally to destroy enemy aircraft in aerial combat, the F-100 later became a fighter-bomber. It made its combat debut during the Vietnam conflict where it was assigned the task of attacking such targets as bridges, river barges, road junctions, and areas being used by infiltrating enemy soldiers. *(Information courtesy of the USAF Museum).*

An F-100A is on display at Buckley AFB.

North American F-100D Super Sabre
(HAS Photo)

The F-100D was an attack version of the Super Sabre with many modifications incorporated into its design. It was also the most extensively built variant. It had a flapped wing, provisions for electronic-countermeasures (ECM) equipment and introduced a low-altitude bombing system (LABS) which made use of a "toss" delivery of nuclear weapons. A total of 1,274 F-100Ds were built. *(The Complete Encyclopedia of World Aircraft).*

The F-100D shown above is on display at the PWAM; another is on display at Buckley AFB. The PWAM's F-100D was flown in by the USAF in Oct. 1972, and later demilitarized by the Colorado Air National Guard. It is on exhibit, courtesy of the USAF. *(Information courtesy of the PWAM).*

North American F-100F Super Sabre

The F-100F is a tandem two-seat training version of the Super Sabre. A total of 339 of the F-100Fs were built. *(The Complete Encyclopedia of World Aircraft)*.

An F-100F is on display at the USAF Academy. This aircraft was flown by Brigadier-General Robinson Reisner in a re-enactment of Charles Lindbergh's historic first solo flight across the Atlantic. Reisner was a POW during the Vietnam War, and retired as a four-star General.

North American T-28C Trojan

The Trojan is a two-seat basic trainer powered by one 1,425-hp Wright Cyclone R-1820-86 radial piston engine. It has a maximum speed of 343 mph, a service ceiling of 35,500', and a range of 1,060 miles. *(The Complete Encyclopedia of World Aircraft)*.

North American Aviation Inc. built the T-28C. This type of aircraft made its first flight on 15 September 1956. It was designed to be used for training of pilots in transition to high-speed aircraft.

Originally produced by North American for the Air Force, the all metal, tricycle landing gear T-28A Trojan made its maiden flight on 26 September 1949. With a top speed of 292 mph, 800-nm range and a service ceiling of 31,650', the Navy began evaluating it as a replacement for the SNJ Texan trainer. The results showed that the Navy version required a more powerful engine and North American responded with the T-28B that was fitted with a 1,425-hp Wright R-1820 engine, which upped the speed to 343 mph. First flown on 6 April 1953, the delivery of 489 T-28As began in 1954, ultimately equipping eleven training squadrons in both the primary and basic training roles.

The success of the T-28Bs resulted in the placement of an order for the tailhook equipped T-28C with a strengthened rear fuselage to absorb the tremendous stress of carrier landings. Making its first flight on 19 September 1955, the T-28C became the center-piece for all

Navy flight training, including basic, primary, instrument, and carrier qualifications.

The T-28 was actively utilized in the Viet Nam war as the T-28D Nomad equipped to carry a variety of weapons ranging from bombs and rockets to napalm and used for counterinsurgency missions throughout Southeast Asia. It was especially effective in night operations against targets not protected by radar controlled anti-aircraft batteries and as armed escorts of A-26 attack aircraft and helicopters. They also operated in hunter/killer teams with O-1 aircraft that used starlight scopes to locate enemy convoys and then call in the T-28s to attack.

On 23 June 1976, the first of the Navy T-28s began to be replaced by the T-34 Mentor. The last was retired from active duty in the spring of 1984, ending twenty-nine years of active service. *(Information courtesy of the National Museum of Naval Aviation).*

The Pueblo Air Museum's T-28C Trojan was accepted for use by the Navy on 16 February 1956 and was used for training. It was released for display by the Department of Navy and flown to Pueblo from Davis Monthan AFB, AZ. *(Information courtesy of the PWAM).*

North American RA-5C Vigilante
(HAS Photo)

The Vigilante is a carrier-based long-range reconnaissance aircraft powered by two 17,860-lb afterburning thrust General Electric J79-GE-10 turbojets. It has a maximum speed of Mach 2.1, a service ceiling of 48,400', and a range of 3,000 miles. *(The Complete Encyclopedia of World Aircraft)*.

In 1955 the US Navy wanted a bomber capable of carrying nuclear weapons at Mach 2. The Vigilante was the Navy's first strategic bomber equivalent to the USAF's contemporary land based aircraft. However, due to the complex bomb launching system, which launched out of tubes at the rear and operational policy changes in the Department of Defense, the Navy designed the second version as a reconnaissance type aircraft which first flew in 1962.

Development of the North American A3J Vigilante began in 1952. Design criteria specified a two-seat, Mach 2 all-weather carrier-based bomber for delivery of nuclear weapons. Fitted with the new low-altitude bombing system (LABS) and an inertial navigation system (INS), the A3J was capable of carrying a 3,020-lb Mk. 27 nuclear store or a 1,885-lb configuration of the Mk-28 weapon. Utilizing such methods as the loft bombing technique, the weapon would be ejected rearward from its internal linear bomb bay. Attached to the weapon were two empty fuel tanks to improve weapon aerodynamics and whose contents had been used en route to the target (a system fraught with problems that were never completely corrected). A high wing layout with a swept wing design, the Vigilante was the first production aircraft to use variable geometry intakes for its two side-by-side engines.

The A3J was first flown in August 1958 and established a new world altitude record of 91,450 feet in December 1963 while carrying a payload of 1,000 kilograms, thus surpassing the existing record by over five miles.

The first squadron deployment occurred in August 1962 aboard the USS Enterprise on its first cruise. Shortly thereafter the Navy's strategic bombing mission was assumed by nuclear powered submarine Polaris missiles. The A3J's mission then reverted to that of photo-reconnaissance with the introduction in 1963 of the RA-5C Vigilante. This version incorporated a sophisticated electronic reconnaissance pod in its bomb bay that included a side-looking airborne radar in a fairing under the fuselage, vertical, oblique and split-image cameras as well as active and passive ECM equipment. Production deliveries began in mid-1964 and shortly thereafter began flying reconnaissance missions over Vietnam from carriers in the Gulf of Tonkin in which nineteen of them were lost to enemy fire.

The INS combined with an automatic flight control system enabled the RA-5C to fly precise courses on mission altitudes ranging from high to tree-top levels. Each photo taken carried a marginal notation that displayed latitude and longitude of the plane at the time it was taken, thus pinpointing target locations. The information obtained was then incorporated into a shipboard data bank and used for mission planning. Several RA-5Cs were also outfitted with a probe-and-drogue refueling package to serve as buddy tankers for compatible aircraft. Despite its Mach 2 capability, several Vigilantes were lost to surface-to-air missiles and/or anti-aircraft fire. Gradual disestablishment of the Vigilante force occurred during 1979-1980 with the deactivation of the last RA-5C squadron, and phasing out of the last of the 156 RA-5/A-3Js produced. *(Information courtesy of the National Museum of Naval Aviation).*

The PWAM's aircraft shown above is the RA-5C Reconnaissance version. *(Information courtesy of the PWAM).*

Northrop X-4 Skylancer

The X-4 was developed for the study of flight characteristics of swept-wing semi-tailless aircraft at transonic speeds (about Mach .85). Northrop built two X-4s. The No.1 aircraft was first flown by Northrop on 16 December 1948, and the second X-4 made its initial flight on 7 June 1949. The No.1 aircraft was grounded after its 10th flight to provide spare parts for the No. 2 aircraft. Northrop's part of the test program ended on 17 February 1950, with the 20th flight of the remaining X-4. Although both aircraft were turned over to the Air Force and then to the National Advisory Committee for Aeronautics (NACA) in May 1950, only the No. 2 X-4 was used in the joint USAF/NACA program to explore stability problems near the speed of sound. The program ended in September 1953 with the 102nd and last flight of the No. 2 aircraft, after proving that swept wing aircraft without horizontal tails were not suitable for transonic flight. Both aircraft survived the test program without serious incident. *(Information courtesy of the USAF Museum).*

The No. 1 X-4 was on display in the lower courtyard at the USAF Academy. It has Captain Chuck Yeager's name stenciled on the left side, and had been at the Academy since 1954. In May 2002 it was returned to Edwards AFB in California. The No. 2 X-4 is on display in the USAF Museum, Dayton, Ohio.

Northrop F-89J Scorpion
(HAS Photo)

The Scorpion is a two-seat, twin-engine, all-weather fighter-interceptor designed to locate, intercept, and destroy enemy aircraft by day or night under all types of weather conditions. It carried a pilot in the forward cockpit and a radar operator in the rear who guided the pilot into the proper attack position. The Scorpion was powered by a pair of after-burning thrust Allison J35-A-35, -33A, -41, or -47 turbojets, which gave it a maximum speed of 636 mph, a service ceiling of 49,200', and a range of 2,600 miles. The Scorpion was armed with 104 70mm rockets in wingtip pods or 27 rockets plus 3 Falcon missiles. *(The Complete Encyclopedia of World Aircraft).*

The first F-89 made its initial flight in August 1948 and deliveries to the Air Force began in July 1950. Northrop produced 1,050 F-89s. On 19 July 1957, a Genie test rocket was fired from an F-89J, the first time in history that an air-to-air rocket with a nuclear warhead was launched and detonated. Three hundred and fifty F-89Ds were converted to J models, which became the Air Defense Command's first fighter-interceptor to carry nuclear armament. *(Information courtesy of the USAF Museum).*

The Scorpion shown above is on display in the PA&SM Air Park. This aircraft was manufactured at Hawthorne, California and delivered to the USAF on 17 November 1954 as an F-89D. This Scorpion is one of 350 F-89Ds converted in June 1957 to the F-89J configuration, and returned to the USAF on 24 September 1957. *(Information courtesy of the PA&SM).*

Northrop T-38A Talon
(HAS Photo)

The Talon is a two-seat tactical fighter and trainer powered by a pair of 5,000-lb after-burning thrust General Electric J-38-GE-5A turbo-jets. It has a maximum speed of 1,083 mph, a service ceiling of 51,880', and a range of 1,543 miles. The F-5E variant is armed with two 20-mm M-39 cannon in the fuselage nose and carries two AIM-9 Sidewinder missiles on wingtip launchers. *(The Complete Encyclopedia of World Aircraft)*.

In the mid-1950s, the USAF required a trainer with higher performance than the T-33 to better prepare student pilots for the latest tactical aircraft that were then coming into service. The aircraft chosen was the T-38A, because it offered high performance with low maintenance and operating costs. The T-38A became the USAF's first supersonic trainer. The T-38A prototype first flew on 10 April 1959, and production continued until 1972. A total of 1,189 T-38As were built. Some were later modified into AT-38Bs with external armament for weapons training purposes.

Jacqueline Cochran set eight performance records in the fall of 1961 flying a production T-38A and in February 1962 a T-38A set four international time-to-climb records. The USAF Thunderbirds used T-38As from 1974 to 1982 because of their economic operation and high performance. Other users of the T-38A include the U.S. Navy in their Top Gun combat simulation program and the National Aeronautics and Space Administration (NASA). *(Information courtesy of the USAF Museum)*.

The T-38 shown above in Thunderbird colors is mounted on a pylon at the USAF Academy. Another in USAF white colors is on display at Weld County Airport.

Piasecki (Vertol) CH-21C Workhorse Helicopter
(HAS Photo)

The H-21 is a troop and cargo transport helicopter powered by one 1,425-hp Wright R-1820-103 Cyclone radial piston engine. It has a maximum speed of 131 mph, a service ceiling of 7,750', and a range of 400 miles. *(The Complete Encyclopedia of World Aircraft).*

Piasecki/Vertol CH-21 "Shawnee is also known as the ""Work Horse," the "Retriever," or the "Mule," but has more often been referred to as the "Flying Banana." The H-21 made its first flight in April 1952. The aircraft was originally designed by Piasecki to transport men and cargo but was later adapted for the rescue of personnel and for assault operations under combat conditions. Normally having a crew of two (pilot and copilot), the H-21 could carry either 20 fully equipped troops or 12 litter patients. In addition to serving with the USAF, the H-21 was supplied to the U.S. Army, the French navy, the Royal Canadian Air Force and the West German Air Force. *(Information courtesy of the USAF Museum).*

The United States Air Force ordered an evaluation and service trials batch of 18 YH-21s from Piasecki Helicopter Corporation in 1949. Following successful trials with these aircraft, 32 H-21As were ordered. Production of the Workhorse continued with the UH-21B, ordered for service with Troop Carrier Command as an assault transport, and delivered from 1953. Production of the UH-21B for the USAF totaled 163. Total UH-21s was 213.

Piasecki developed the CH-21 arctic rescue helicopter from the world's first tandem rotor transport helicopter, the XHRP-1. This helicopter was first flown in March 1945. The CH-21 was equipped with an R1820-103 Wright Cyclone piston engine de-rated to 1,150 horsepower. It was first flown in April 1952. The CH-21B was equipped with a 1,425 horsepower engine developing full shaft horsepower. The CH-21 was first delivered to the United States Air Force in 1953 and immediately placed into service in the Arctic without any formal service evaluation. The CH-21 had winterization features permitting operation at temperatures as low as -65 degrees Fahrenheit and could be routinely maintenance under severe conditions. Both Canadian and USAF forces in the northern regions of Alaska, Canada, and Greenland in support of Distant Early Warning (DEW) Line installations therefore used it. *(Information courtesy of the Strategic Air and Space Museum)*.

The Pueblo Historical Aircraft Society crew restoration members restored the PWA Museum's aircraft shown above over a period of several years, as parts became available. The crew headed by Ted Baer is composed of Gerald Putnam, Ken Darr, Bob Dyleski & Joe Harmon, all of Pueblo, CO. *(Information courtesy of the PWAM)*.

A CH-21 is also on display inside the WORA&SM.

Piper J-3 Cub/L-4A/O-59 Grasshopper

The L-4A liaison aircraft, originally designated the O-59, was the military version of the famous Piper J3 Cub. The Grasshopper was a two-seat light plane powered by a 65-hp Continental A65 flat-four piston engine. It had a maximum speed of 92 mph, a service ceiling of 12,000', and a range of 250 miles. *(The Complete Encyclopedia of World Aircraft)*.

The Army ordered the first O-59s in 1941 for tests in conjunction with its growing interest in the use of light aircraft for liaison and observation duties in direct support of ground forces. Between 1941 and 1945, the Army procured almost 6,000 Piper Aircraft. During WW II, Grasshoppers performed a wide variety of functions throughout the world such as for artillery fire direction, pilot training, glider pilot instruction, courier service and front-line liaison. *(Information courtesy of the USAF Museum).*

Republic P-47N Thunderbolt

The P-47N Thunderbolt is a single-seat fighter-bomber powered by one 2,300-hp Pratt & Whitney R-2800-21W radial piston engine with water injection. The aircraft had a maximum speed of 697 km/h, a service ceiling of 41,000', and a range of 3058 km. The Thunderbolt was armed with eight .50 cal machine-guns and could carry a bomb-load of 2,500-lbs. *(The Complete Encyclopedia of World Aircraft).*

Affectionately nicknamed "Jug," the P-47 was one of the most famous AAF fighter planes of WW II. Although originally conceived as a lightweight interceptor, the P-47 developed as a heavyweight fighter and made its first flight on May 6, 1941. The first production model was delivered to the AAF in March 1942, and in April 1943 the Thunderbolt flew its first combat mission, a sweep over Western Europe. Used as both a high-altitude escort fighter and a low-level fighter-bomber, the P-47 quickly gained a reputation for ruggedness. Its sturdy construction and air-cooled radial engine enabled the Thunderbolt to absorb severe battle damage and keep flying. During WW II, the P-47 served in almost every active war theater and in the forces of several Allied nations. By the end of WW II, more than 15,600 Thunderbolts had been built. Production P-47B, C, early D and G series aircraft were built with metal-framed "greenhouse" type cockpit canopies. Late D series (dash 25 and later) aircraft, and all M and N series production aircraft were given clear "bubble" canopies, which gave the pilot improved rearward vision. *(Information courtesy of the USAF Museum).*

The P-47N formerly mounted in front of NORAD HQ at Peterson Air Force Base was removed from its pedestal on 16 July 2000 after being mounted there for 29 years. It is presently undergoing restoration inside one of the original Peterson AFB hangars, part of the PA&SM. P-47N (Serial No. 44-89425) was manufactured by Republic Aviation Corporation at Farmingdale, New York, and was delivered to the USAAF on 11 October 1945. It served with fighter units in Indiana, Kansas, Washington, Utah, Maine, Pennsylvania and Oklahoma before being placed in storage in December 1951. It was sent to the Texas Engineering and Manufacturing Corp., Hensley Field TX, for work. On 10 June 1952 it went to the 198th Fighter Squadron (ANG), Puerto Rico, where it served until returned to storage on 22 December 1954. It was on display at Perrin AFB, Texas where it stood as a gate guard in camouflage colors until 1970. It was later delivered to Peterson AFB where it was painted in the colors of a typical P-47N Thunderbolt in service with Aerospace Defense Command (ADC), and then mounted on a pedestal for display in May 1971. *(Information courtesy of the PA&SM).*

Republic F-84C Thunderjet
(HAS Photo)

The Thunderjet was the last sub-sonic straight-wing fighter-bomber used on operations, serving a valuable role in the Korean War. It was powered by a Wright J-65W-3 turbojet, and was armed with six .50 cal Browning M3 machine-guns and could also carry up to 6,000 pounds of external stores.

Large numbers of Thunderjets were built for NATO forces. The F-84, the USAF's first post-war fighter, made its initial flight on the 26th of February 1946. It began rolling off the production lines in June 1947, and by the time production ceased in 1953, approximately 4,450 "straight-wing" F-84s (in contrast to the swept-wing F-84F) had been built. In addition to being used by the USAF, many were supplied to allied nations participating in the Mutual Security Program. During its service life, the F-84 became the first USAF jet fighter able to carry a tactical atomic weapon. The airplane gained its greatest renown during the Korean War where it was used primarily for low-level interdiction missions. Almost daily the F-84 attacked enemy railroads, bridges, supply depots and troop concentrations with bombs, rockets and napalm. (*Information courtesy of the USAF Museum*).

The "Thunderjet" was America's first new fighter after WWII. It also served as escort for B-29 bombers during the Korean War. Later models were equipped with ejection seats and under-wing retractable rocket launchers. The PWAM's F-84C project shown above was donated by Mullen Preparatory High School of Denver CO. It is in poor condition with many parts missing and as time allows, will hopefully be restored to aesthetic standards. (*Information courtesy of the PWAM*).

Republic RF-84K Thunderflash
(HAS Photo)

The swept-wing F-84F evolved from the straight-wing F-84. The prototype first flew on June 3, 1950 and deliveries began in 1954, primarily to the tactical Air Command as a ground support fighter-bomber. Republic built 2,112 "F's" while General Motors fabricated an additional 599. Of these, 1,301 were delivered to NATO air forces. Production of a reconnaissance version, the RF-84F, totaled 718 aircraft, including 386 for allied countries. The RF-84F featured engine air intakes at the wing roots plus cameras in the nose. F-84Fs gradually were replaced by supersonic F-100s in the late 1950s and were turned over to Air National Guard units. However, some F-84Fs were called back to temporary USAF service in the early 1960s due to the Berlin Crisis. *(Information courtesy of the USAF Museum).*

The WORA&SM RF-84K shown above is a rare model. Only 25 K models were equipped with a hook mounted forward of the cockpit for attachment to a "mother-ship" such as a B-36 Pacemaker. After hooking up to their long-range transport, these aircraft were carried by it to the designated reconnaissance area. After being launched, the Thunderflash would conduct its reconnaissance task, and then hook onto its transport again to be brought back to base. *(Information courtesy of the WORA&SM)*.

Republic F-105D Thunderchief
(HAS Photo)

The F-105 evolved from a project begun in 1951 by Republic Aviation to develop a supersonic tactical fighter-bomber to replace the F-84F. The prototype first flew on October 22, 1955, but the first production aircraft, an F-105B, was not delivered to the USAF until 1958. The F-105D all-weather strike fighter and the two-place F-105F dual-purpose trainer-fighter also were built before F-105 production (833 Aircraft). ended in 1964. No "C" or "E" series were produced, and the "G's" were modified from F-105Fs. The F-105D could carry over 12,000 pounds of ordnance, a heavier bomb load than a World War II B-17. Up to 8,000 pounds could be carried internally in the bomb bay. The F-105D was used extensively in the Viet Nam War. The last F-105D was withdrawn from USAF service on July 12, 1980. *(Information courtesy of the USAF Museum)*.

The Thunderchief was the first supersonic fighter-bomber developed from scratch. It was operational from 1960 to 1970, and between 1965 and 1968, 75% of the USAF's assault missions over North Vietnam were carried out by Thunderchiefs. Nicknamed the "Thud," this supersonic fighter-bomber carried over six tons of weaponry and cruised at Mach 2. The aircraft has a coke-bottle-shaped fuselage, narrowing toward the centre, then widening again. This shape, devised at the National Advisory Committee for Aeronautics (NACA), helped minimize drag as the plane accelerated to supersonic speed.

The aircraft shown above on display at the USAF Academy was assembled at McClellan AFB, California, from parts of at least ten sister aircraft that saw combat duty in Southeast Asia. It has been at the Academy since 1968. The Academy aircraft is maintained by the "2nd Cadet Group." *(Information courtesy of the USAF Academy).*

A Thunderchief in silver colors is on display inside the WORA&SM.

Rockwell B-1A Lancer

The B-1 is a "swing-wing" strategic bomber of blended wing-body design intended for high-speed low-altitude penetration missions. It is powered by four General Electric 010-GE-102 turbofans. The aircraft has three internal weapons bays and can carry up to 84 500-pound bombs or eight air-launched cruise missiles or other weapons. Eventually, it will replace the aging B-52 bomber, operational since the 1950s. The B-1 uses shorter runways than the B-52, can carry twice the payload, and has a smaller radar profile than the B-52's, making it harder for the enemy to detect.

Construction of the first prototype B-1A began in late 1972 and the first flight occurred on 23 December 1974. By the end of June 1977, three B-1As had made 118 flights totaling 646 hours of flying time with over 21 hours at supersonic speed and more than 35 hours at high speed below 500 ft. B-1A production was canceled by presidential decision on 30 June 1977, but in January 1982, as a result of another

Presidential decision, the USAF directed Rockwell International to begin production of 100 B-1Bs, an improved version of the B-1A. *(Information courtesy of the USAF Museum).*

An original prototype B-1 Lancer is on display in the WORA&SM.

Ryan ST-A/PT-22 Recruit

The PT-22 Recruit a primary trainer powered by one 160-hp Kinner R-540-1 radial piston engine. It had a maximum speed of 131 mph, a service ceiling of 15,500', and a range of 352 miles. *(The Complete Encyclopedia of World Aircraft).*

Primary trainers represented the first of three stages of military flight training, including primary, basic and advanced. Prior to 1939, the Air Corps relied entirely on biplanes as primary trainers, but in 1940 it ordered a small number of Ryan low-wing civilian trainers and designated them as PT-16s. They were so successful that the Air Corps then ordered large numbers of improved versions, among them the PT-22. By the time production was completed in 1942, 1,023 PT-22s had been delivered. Twenty-five additional trainers, ordered for the Netherlands, were taken over by the Air Corps in 1942 and designated as PT-22As. *(Information courtesy of the USAF Museum)*

In response to a need for mass expansion of aviation training programs in both the Navy and the Army, companies such as Ryan, Stearman, and Vultee were contracted to produce standardized primary and basic trainers for both services. The Ryan Aeronautical Corporation's design (the ST-3) was the first ever monoplane purchased by the Army Air Corps in 1939 as a primary trainer. One hundred of them were purchased by the Navy the following year, redesignated as the NR-1 Recruit, and served the Naval Air Training Command until 1944.

Total production of NR-1/ST-3 aircraft amounted to 1,023 models with the last one being produced in 1942. It was at this time that Ryan was asked to investigate production possibilities of a version that contained no strategic materials. What followed was a virtually new airplane design constructed of plastic-bonded plywood of which the

Army ordered five experimental models. *(Information courtesy of the National Museum of Naval Aviation).*

Schweitzer TG-3A Glider

The TG-3A is a two-place, dual-control glider manufactured for the Army Air Forces by the Schweitzer Aircraft Co. during WW II. It was derived from the Schweitzer SGS 2-12 high-performance sailplane (designated XTG-3 by the AAF). The AAF evaluated the aircraft in 1942 for suitability as a trainer for cargo glider pilots. After testing three XTG-3s, the AAF ordered 110 TG-3As from Schweitzer (50 more were ordered from Air Glider, Inc. but only one was built). Student glider pilots normally received about six hour's dual instruction in the TG-3As before being trained in the large CG-4A cargo glider.

The TG-3A's wings are made of spruce and mahogany plywood covered with doped fabric. It is equipped with spoilers (used to increase descent rate) as well as ailerons. The fuselage is constructed of welded chrome-molybdenum steel tubing, which is fabric covered. The landing gear is a single unsprung wheel with a disc brake. In addition, there are skids on the nose, tail, and wing tips. *(Information courtesy of the USAF Museum).*

A TG-3A is on display suspended from the ceiling inside the Visitor's Center at the USAF Academy in Colorado Springs.

Sikorsky S-58 Choctaw/SH-34J Sea Horse
(HAS Photo)

Sikorsky started to design the S-58 in 1951 at the request of the US Navy, which was looking for a new anti-submarine helicopter to replace the Sikorsky S-55, deemed to be too small and underpowered. The S-58 first flew on 8 March 1954 and the US production run amounted to 1821. The Navy version was designated HSS 1 and the army version H34. Both versions were exported in numerous countries (Germany, Belgium, Italy, Canada, etc.).

For the war in North Africa, the French armed forces increased their rotorcraft fleets dramatically. 90 S-58's were bought from Sikorsky and 171 were built under license by Sud Aviation. The French navy operated the HSS between 1958 and 1978 for transport and anti-submarine warfare. They were replaced by the Westland Lynx in this latter mission. The French air force flew the H 34 for transport and rescue between 1958 and 1984, when they were replaced by the Aerospatiale Puma. In North Africa, the S-58 demonstrated the helicopter gunship concept, providing firepower during the landing of the troop transport helicopters. This method was also applied by the USA in Vietnam and is now in worldwide use.

The S-58 was also used to carry airline passengers between airport and city center both in the USA and Europe. Needless to say that the cabin fit was very different from the military version. Westland also bought the S-58 license from Sikorsky and built 352 Wessexes, all powered by turboshafts housed in a modified nose. The Wessex is still in use with the British armed forces. In 1970, Sikorsky did the same kind of modification to an S-58, replacing the piston engine by two PW Canada PT 6 turboshafts in a modified nose. These conversions were designated S-58 T and several are still flying worldwide; whereas the piston engined version has almost disappeared.

The Pueblo Air Museum's SH-34J shown above was donated in 1980 by the US Naval Air Systems Command in Washington D.C. and was transported by truck from Davis Monthan AFB, AZ. *(Information courtesy of the PWAM)*.

Other H-34s are on display at Fort Carson and at Buckley AFB.

Sopwith Pup

The Sopwith Pup is a copy of a single-seat biplane fighter aircraft manufactured in 1916. It has a wingspan of 26.5', and a length of 19.31'. The original Pup was powered by one 80-hp Le Rhône rotary piston engine. It had a maximum speed of 112-mph, a service ceiling of 17,500' and a rate of endurance of 3 hours. It was armed with one forward-firing synchronized .303-in Vickers machine-gun and could carry up to four 25-lb bombs on external racks. *(The Complete Encyclopedia of World Aircraft)*.

The Sopwith Pup ranks as one of the truly great combat aircraft of World War I. Some have called it the most perfect flying machine ever made. Regardless, the Pup was undeniably a docile and enjoyable airplane to fly. A true pursuit, the Pup was capable of competing one-on-one with any air combat aircraft in the sky at the time of its debut during the spring of 1916. It proved so effective against German aircraft, German pilots consciously avoided confrontations with it until the advent of more capable German pursuits. By the end of 1917, the Pup's advantage had been offset by newer designs. It was quickly phased-out as more advanced aircraft became available.

The Pup was equipped with a single-fixed, synchronized Vickers machine gun mounted on top of the forward fuselage, ahead of the cockpit. *(Information courtesy of the Museum of Flight)*

Supermarine 361 Spitfire Mk. IX

The Spitfire Mk. IX is a single-seat interceptor fighter powered by one 2,050-hp Griffon 65 or 66 piston engine with a five-bladed propeller. It had a maximum speed of 369 mph, a service ceiling of 36,500', and a range of 1,135 miles. The Mk. IX variant had a broad tail and in some cases, a tear-drop canopy. Its wings were a standard "F" type and it was armed with eight .303-in Browning machine-guns or could be armed with four .303-in machine-guns and two 20-mm cannon.

Unquestionably the most important British fighter of World War II, and not inconsequentially one of the most important piston-engined fighters of all time, Supermarine's immortal Spitfire came to life during the mid-1930s as a result of the successes the company had enjoyed in major air race events, most notably the Schneider Cup competitions of the 1920s and 1930s. The first Spitfire flew in 1936 and by the beginning of World War II the type was in limited production.

Early Spitfires met their match in the Me 109, and later, the Focke-Wulf Fw 190, but steady improvements in the airframe and engine eventually created a fighter that was the equal of anything the Axis could throw into the sky.

Perhaps the most successful of the many Spitfire variants was the Mk. IX. Supermarine built over 5,700 of this particular version. It quickly became the most important British fighter of World War II. *(Information courtesy of the Museum of Flight)*.

Bill Greenwood, Aspen, CO flies a Supermarine Spitfire Mk. Tr.IXc (Serial No. TE308), Reg. No. N308WK. This aircraft was built at Castle Bromwich, UK. It does not have a record of service with the RAF, but was sold to Vickers Aviation as non-effective on 19 July 1950. This Spitfire was converted to a 2-seater with registration number G-AWGB. It saw service with the Irish Air Corps as No. 163 on 10

July 1951, and was struck off charge in 1961. It was used in the film "Battle of Britain" in 1968. It became CF-RAF in 1974, Reg. No. N92477, before going to Aspen, CO with Reg. No. N308WK.

Vought F-8U Crusader
(HAS Photo)

The Crusader is a super sonic air superiority fighter designed for operation from the USN's aircraft carriers. The Crusader was powered by one 18,000-lb thrust Pratt & Whitney J57-P-20A turbojet, and had a maximum speed of 1,120 mph, a service ceiling of 58,000', and a range of 1,000 miles. This Navy and Marine fighter aircraft was equipped with search and fire-control radar, was armed with four 20mm Colt-Browning cannon, two Sidewinder missiles and an under-fuselage rocket pack and or other weapons. The fighter was equipped with search and fire-control radar, and could carry air-to-air or air-to-surface missiles, bombs and rockets. *(The Complete Encyclopedia of World Aircraft)*.

During the first period of the air war in Vietnam, from 1965 to 1968, USN pilots flying the Crusader compiled the best victory ratio of any aircraft in the war: six downed enemy aircraft for every Crusader lost. It was the basis for the design of the A-7 Corsair II. With its high speed and ability to perform at above 95% of the earth's atmosphere it was designed for defense against a Mach-2 bomber. It had a tilting 2-

position wing providing for carrier use. It was adapted to carry special weapons and electronic intelligence equipment for photo-reconnaissance. It also carried an external fuel tank. . *(Information courtesy of the National Museum of Naval Aviation)*.

The Crusader shown above is on display at the PWAM. It was built in 1958 and donated to the city of Pueblo by the Mullen Preparatory High School of Denver CO. *(Information courtesy of the PWAM)*.

Vought A-7D Corsair II
(HAS Photo)

The A-7D Corsair II is a carrier-based attack bomber powered by a 14,500-lb thrust Allison TF41-A-2 turbofan, or by a Pratt & Whitney J57-P-20A turbojet. It has a maximum speed of 698 mph, and a tactical radius with typical weapon load of 700 miles. It is armed with one 20mm M61A1 multi-barreled cannon, two Sidewinder missiles and an under-fuselage rocket pack. It can also carry 15,000-lbs of mixed stores externally. *(The Complete Encyclopedia of World Aircraft)*.

The Ling-Temco-Vought design consortium developed the Corsair II based on the F-8 Crusader. The Corsair II first entered service with training squadrons in 1966. The aircraft is a sub-sonic machine, which can carry up to 20,000 pounds of offensive weapons. The A-7D is a single-seat, tactical close air support aircraft. Although designed primarily as a ground attack aircraft, it also has limited air-to-air combat capability. It was derived from the basic A-7 originally developed by LTV for the U.S. Navy.

The first A-7D made its initial flight on 5 April 1968, and deliveries of production models began on Dec. 23, 1968. When A-7D production ended in 1976, 459 had been delivered to the USAF. In 1973, the USAF began assigning A-7Ds to the Air National Guard (ANG), and by 1987 they were being flown by ANG units in ten states and Puerto Rico. The A-7D demonstrated its outstanding capability to attack ground targets while flown by the 354th Tactical Fighter Wing at Korat RTAFB, Thailand, during the closing months of the war in Southeast Asia. The Corsair II achieved its excellent accuracy with the aid of an automatic electronic navigation and weapon delivery system. *(Information courtesy of the USAF Museum).*

The Corsair II shown above is on display at Buckley AFB. A US Navy Corsair II is on display at the Naval Reserve Component at Buckley AFB. Others may be viewed in the WORA&SM; at the Colorado Air National Guard unit in Greeley, and at Montrose airport.

At the entrance to the Montrose airport along U.S. Highway 50 is a Corsair A-7 which Montrose is proud to have gotten from the Colorado National Guard. The Corsair, nicknamed the "Double Nickel" because of its official number (055) won the 1981 "Gunsmoke" competition at Nellis Air Force Base, while being piloted by "Top Gun" Lt. Col. Wayne Schultz. The competition tests the best units in the U.S. Air Force, Air Force Reserve and Air National Guard. Colorado won the Top Team award in 1981.

Yakovlev Yak-11

The Yak-11 is an advanced trainer/liaison aircraft powered by one 570-hp Shvetsov Ash-21 radial piston engine. It has a top speed of 289 mph, a service ceiling of 23,295', and a range of 795 miles. It could be armed with one synchronized .5-in UBS or .303-in ShKAS machine-gun. This two-seat trainer was first flown in the USSR in 1945. (*The Complete Encyclopedia of World Aircraft*).

U.S. Army Hawk Missile

The Hawk is a medium range, surface-to-air guided missile that provides air defense coverage against low-to-medium-altitude aircraft. It is a mobile, all-weather day and night system. The missile is highly lethal, reliable, and effective against electronic countermeasures. The basic Hawk was developed in the 1950s and initially fielded in 1960. The system has been upgraded through a series of product improvements beginning with the Improved Hawk in 1970. The Phase III product improvement and the latest missile modification were first fielded in the early 1990s to the U.S. Army and U.S. Marine Corps.

Two missile modifications have extended the missile's field life and added electronic counter-countermeasures to defeat special threats. Development was completed on Hawk mobility, the latest system improvement, and it was produced for the USMC and Sweden. The U.S. Army National Guard (ARNG) as well as 20 allied nations, including NATO and several countries in both Southwest Asia and Southeast Asia, were also equipped with the Hawk system. Prime contractor for this system is the Raytheon Company. The cost per missile is $250,000; per fire unit, $15 million; and per battery, $30 million.

Although the U.S. Army deployed Hawk missile batteries during the conflicts in Vietnam and Persian Gulf, American troops have never fired this weapon in combat. The first combat use of Hawk occurred in 1967 when Israel successfully fired the missiles during the Six-Day War with Egypt. Even though it was not used by the coalition during Operation Desert Storm, the Hawk missile did see action during the Persian Gulf War. Kuwaiti air defense units equipped with U.S. Hawk antiaircraft missiles downed about 22 Iraqi aircraft and one combat helicopter during the invasion of 2 August 1990. The system later posed a possible threat to the U.S.-led coalition because Iraqi forces captured both Hawk and TOW missiles in Kuwait. *(Internet: www.redstone.army.mil/history/systems/HAWK).*

A Hawk missile system is on display at the PA&SM.

U.S. Army MIM-3 Nike Ajax Missile

The Nike Ajax is a slim two-stage guided missile powered by a liquid-fueled motor utilizing a combination of inhibited red fuming nitric acid (IRFNA), asymmetrical dimethyl hydrazine (UDMH) and JP-4 jet petroleum. The Ajax was developed in 1951, and it is launched form a fixed platform with a jettisonable solid fuel rocket booster which burns for about 3 seconds, accelerating the missile with a power of 25 times the force of gravity. The Ajax missile was capable of maximum speeds of over 1,600-mph and could reach targets at altitudes of up to 70,000 feet, although it had a limited range of roughly 25 miles. The missile was armed with three individual high-explosive, fragmentation-type warheads located at the front, center and rear of the missile body. The Nike Ajax was replaced by the Nike Hercules. *(Information courtesy of the PA&SM).*

An Ajax missile is on display at the PA&SM.

U.S. Army MIM-14 Nike Hercules Missile

The Nike Hercules is a medium-range, ground-based, solid-propellant theater defense missile. It was first developed in 1954 and introduced to the U.S. Army in 1958, replacing the Nike Ajax. It is a two-stage missile with a tandem-mountable jettisonable cluster boost motor. The missile has a range of 145 km and an engagement envelope of between 1,000 and 25,000 meters. It works in conjunction with a high-power acquisition radar (HIPAR). Phased out of service in the 1980's, 25,000 were produced, with 2,650 being exported. *(Information courtesy of the PA&SM).*

A Nike Hercules is on display at Peterson AFB.

Minuteman II Missile

(HAS Photo)

The Minuteman is a solid-fuel, intercontinental-range ballistic missile capable of being fired from silo launchers and delivering a thermonuclear payload of one or several warheads with high accuracy over great distances. The Minuteman III intercontinental ballistic missile (ICBM) is the most advanced version of the solid-propellant series of weapons and offers greater range than the Minuteman I and II. Its larger nuclear payload consists of multiple independently-targeted reentry vehicles (MIRV) which, with such aids as chaff and decoys, increase its chances of penetrating enemy defenses.

The Minuteman I became operational in 1962 and three years later, Minuteman II reached the same status. Minuteman I sites were later modified to accept the improved -II and -III versions, permitting the updating of the entire Minuteman force of approximately 1,000 missiles scattered at launch sites in central and northern plains states.

On 19 June 1970, the Strategic Air Command at Minot AFB, North Dakota, accepted the first operational Minuteman IIIs. A flight of ten launchers has a launch control center located approximately 50 feet underground. To prevent an unauthorized launch, it requires the coordinated efforts of two-man teams of SAC launch control officers to fire one of these missiles skyward from its hardened underground silo. *(Information courtesy of the USAF Museum).*

The Minute Man II shown above is on display at the USAF Academy in Colorado Springs.

Epilogue

For all those who serve in the defense of North America, particularly since the terrorist attack on the World Trade Center and the Pentagon on 11 September 2001 and in the ongoing war against terrorism, we thank you.

The intention of this book is to record and identify the locations of where at least some those aircraft which have been used in that defense can be found in Colorado. They are, in fact, Colorado Warbird Survivors.

Afterword

There is a far better way to protect our homes and our people than to fight and win a great war. The better way is to be so obviously superior in our ability to carry the war to an enemy that he will not take the risk of starting one.

Chief of Staff General Thomas D. White.

APPENDIX A

Short List of Colorado Warbird Survivors 2003

1. Alexander Eaglerock Model A-14

2. Avro CF-100 Mk. 5C Canuck

3. Beech C-45F Expeditor

4. Beech C-45G Expeditor

5. Beech C-45H Expeditor

6. Bell Model 47 UH-13J Sioux Helicopter

7. Bell Model 204 HU-1 Iroquois (Huey) Helicopter

8. Bell Model 209 AH-1F Huey Cobra Attack Helicopter

9. Bell Model 209 AH-1G Huey Cobra Attack Helicopter

10. Bell OH-58C Kiowa

11. Boeing/Stearman Model 75/PT-13D/N2S-5 Kaydet

12. Boeing B-29A Superfortress

13. Boeing Model 367 KC-97L Stratotanker

14. Boeing B-47E Stratojet

15. Boeing B-52D Stratofortress

16. Boeing CIM-10A BOMARC Missile

17. Cessna T-41A Mescalero

18. Cessna O-2 Skymaster

19. Cessna Model 310 U-3 Administrator "Blue Canoe"

20. Chance Vought V-346 F7U-3 Cutlass

21. Convair F-102A Delta Dagger

22. Convair F-106A Delta Dart

23. Convair HC-131A Samaritan

24. Curtiss JN-4D Jenny

25. Curtiss P-40E Warhawk

26. Douglas B-18A Bolo

27. Douglas DC-3/C-47 Skytrain

28. Douglas AD-4NA Skyraider

29. Douglas A-26C Invader

30. Douglas B-26C Invader

31. Douglas F-6A Skyray

32. Fairchild C-119 Boxcar

33. Fairchild Republic A-10A Thunderbolt II

34. Fokker D.VII

35. Fokker Dr.I

36. Folland Fo.141 Gnat

37. General Dynamics FB-111A Aardvark

38. General Dynamics F-16A Fighting Falcon

39. Goodyear FG-1D Corsair

40. Grumman G-44A/J4F-1 Widgeon

41. Grumman TBM-3E Avenger

42. Grumman HU-16C Albatross

43. Grumman F9F-8 Cougar

44. Grumman F11F-1 Tiger

45. Grumman A-6E Intruder

46. Grumman OV-1D Mohawk

47. Grumman F-14A Tomcat

48. Hiller H-23D Raven

49. Hawker Sea Fury T Mk. 20

50. Hunting (Percival) P.84 Jet Provost

51. Lockheed RB-37 Ventura

52. Lockheed Model 18/C-60A-5 Lodestar

53. Lockheed P-38L (F-5G) Lightning

54. Lockheed P-38L-5 Lightning

55. Lockheed 1049C Constellation

56. Lockheed P-80 Shooting Star

57. Lockheed T-33A Shooting Star

58. Lockheed F-94C Starfire

59. Lockheed F-104C Starfighter

60. Lockheed EC-121T Warning Star

61. Lockheed P2V-5 Neptune

62. Martin EB-57E Canberra

63. Martin Marietta SV-5J (X-24A) Lifting Body

64. McDonnell Douglas A-4D-2 Skyhawk

65. McDonnell F-101B Voodoo

66. Mcdonnell Douglas F-4C Phantom

67. Mcdonnell Douglas F-4E Phantom

68. McDonnell Douglas F-15A Eagle

69. Mikoyan Gurevich MiG-17F/JJ-5

70. Nord 3202 L'Armée d'Air

71. North American AT-6A Texan

72. North American AT-6C Texan

73. North American AT-6D Texan

74. North American AT-6F Texan

75. North American AT-6G Texan

76. North American P-51D Mustang

77. North American F-51D Mustang

78. North American F-86F Sabre

79. North American F-86H Sabre

80. North American F-86L Sabre Dog

81. North American F-100A Super Sabre

82. North American F-100D Super Sabre

83. North American F-100F Super Sabre

84. North American T-28C Trojan

85. North American RA-5C Vigilante

86. Northrop X-4 Skylancer

87. Northrop F-89J Scorpion

88. Northrop T-38A Talon

89. Piasecki CH-21C Workhorse Helicopter

90. Piper J-3 Cub

91. Republic P-47N Thunderbolt

92. Republic F-84C Thunderjet

93. Republic RF-84K Thunderflash

94. Republic F-105D Thunderchief

95. Rockwell B-1A Lancer

96. Royal Aircraft Factory SE-5A

97. Schweitzer Glider

98. Sikorsky SH-34J/S-58 Choctaw

99. Sopwith Pup

100. Supermarine Spitfire Mk. IX

101. Vought F-8U Crusader

102. Vought A-7D Corsair II

103. U.S. Army Hawk Missile

104. U.S. Army MIM-3 Nike Ajax Missile

105. U.S. Army MIM-14 Nike Hercules Missile

106. Minuteman II Missile

APPENDIX B

History of Peterson Air Force Base

Information courtesy of the 21st Space Wing Public Affairs office and the Peterson Air and Space Museum.

Peterson AFB traces its roots to the Colorado Springs Army Air Base, established on 6 May 1942 at the Colorado Springs Municipal Airport which had been in operation since 1926. The base carried out photo-reconnaissance training under the auspices of the Photo Reconnaissance Operational Training Unit (PROTU). On 22 June 1942 Colorado Springs Army Air Base was assigned to the 2nd Air Force, headquartered at Fort George Wright, Washington.

Then, after only a few weeks, a tragedy occurred that would indelibly affect the base. On 8 August 1942, First Lieutenant Edward J. Peterson, Operations Officer for the 14th Photo Reconnaissance Squadron and a native of Colorado, crashed while attempting to take off from the airfield when the left engine of his twin engine F4 (a reconnaissance variant of the P-38 Lightning) failed. A base fire department crew rescued Lt. Peterson from the burning wreckage. Unfortunately, Lt. Peterson sustained significant burns and died at a local hospital that afternoon, thereby becoming the first Coloradan killed in a flying accident at the airfield. Consequently, on 13 December 1942, officials changed the name of the Colorado Springs Army Air Base to Peterson Army Air Base in honor of the fallen airman.

The base assumed a new mission in the spring of 1943, that of heavy bomber combat crew training. The 214th Combat Crew Training School conducted the training, utilizing the B-24 Liberator. From 5 March to 1 October 1943, "Peterson Field" as the base was commonly called, was assigned to the 3rd Air Force, headquartered at Greenville Army Air Base, South Carolina. Control of Peterson Field then reverted to the 2nd Air Force. In June 1944 the mission at the base once again changed, this time to fighter pilot training. The 72nd Fighter Wing, assigned to the base, employed P-40 Warhawks to carry out this mission.

In April 1945, Peterson Field was assigned to Continental Air Forces. The location of the Army Air Forces Instructors School at the base signaled another mission change. A short time later, on 31 December 1945, the Army inactivated the base, turning the property over to the City of Colorado Springs.

The legacy of Peterson Field and the military presence in Colorado Springs took a significant turn in September 1947, following the birth of the United States Air Force. Soon after its inception, the fledgling service twice reactivated the base, from 29 September 1947 to 15 January 1948 and again from 22 September 1948 into 1949. During the latter period, the base served as an airfield for Headquarters, 15th Air Force which had been temporarily located in Colorado Springs. Peterson Field inactivated again when 15th Air Force moved to March Air Force Base in 1949.

The Air Force activated Peterson Field once more, following the January 1951 establishment of Air Defense Command at Ent AFB, located in downtown Colorado Springs. The 4600th Air Base Group activated simultaneously on 1 January 1951 and provided support for the newly established command. In 1958 the 4600th achieved wing status and was designated as the 4600th Air Base Wing. Subsequently, on 1 April 1975, the Air Force redesignated the wing as the 46th Aerospace Defense Wing. One year later, on 1 March 1976, Peterson Field was renamed Peterson Air Force Base.

Strategic Air Command assumed control of the base on 1 October 1979. Then, on 1 September 1982, USAF officials activated Air Force Space Command at Peterson, followed by the activation of the 1st Space Wing on 1 January 1983. Peterson Air Force Base became the hub of Air Forces space activity when the 1st Space Wing assumed host unit responsibility following the inactivation of the 46th Aerospace Defense Wing on 1 April 1983. The 1st Space Wing then transferred host unit responsibility to the 3rd Space Support Wing which activated on 15 October 1986. Finally, on 15 May 1992, these two wings inactivated and their personnel and equipment transferred to the 21st Space Wing which activated on 15 May 1992.

APPENDIX C

The Lafayette Escadrille—Remembering Dr. James Parks

Ariticle by Jeff Price and Di Freeze

When Andy Parks was six, his father, Jim Parks, decided to build a replica of a Fokker D.VII, a World War I biplane. When Andy was 37, just two years ago, at an event in his father's honor, he said that building the plane, "Aquarius," was a "bonding phase" for he and his father. Andy Parks had a lot of valuable time to bond with his father, before he passed away on 22 August 2002, after being with diagnosed with Alzheimer's 13 years ago. Many others did as well. James "Jim" John Parks was born 30 January 1929. When he was still in short pants, he took his first airplane ride with famed South Dakota aviator Clyde Ice.

Jim's dad, Fred Parks, served in World War I as an infantryman. He was set to be reassigned to the air service when he was gassed in the Muesse Argon Offensive, and left with 50 percent of his total air capacity in one lung and 40 percent in the other. The incident stopped any chances of him flying for the military. After the war, he went on to a professional boxing career, with 29 bouts and 26 first round knockouts. When the military told him he was too old for WWII, Grandfather Parks remembered an earlier promise he had made to God on the battlefield, after being gassed.

"He was a religious man; he promised God that if He got him through that, he would become a priest," says Andy Parks. Grandfather

167

Parks entered the seminary, as there wasn't an age limit on priests entering the war.

During the thirties, Jim's parents operated a boarding house in Rapid City, S.D. Visitors included pilots and mechanics, who, all hours of the day and night, would sit around the dinner table and tell war stories.

Those visitors included Ken Porter, an ace with the 147th Aero Squadron, and Doug Campbell, who after dropping out of Harvard University, enlisted in the U.S. Air Service. Assigned to the 94th Aero Squadron in France, he and Alan Winslow shared credit for the 94th Aero Squadron's first official victory over an enemy aircraft. Flying a Nieuport 28, Campbell was the first U.S. Air Service pilot trained in the United States to score five confirmed victories, becoming an ace, beating Eddie Rickenbacker by one day.

Both these men became good friends of Jim, as did George Vaughn, who had 19 victories, and Ray Brooks, whose Spad aircraft hangs in the Smithsonian. One of these visitors was Frank Smith, a mechanic with the United States in WWI, who told a story that James found intriguing. Smith was assigned as a balloon rigger in 1918. One day, he found a rigging error on the plane of a French pilot. He left a "red tag" on the joystick, telling him not to fly until they fixed the problem. When the pilot found out, he knew that Smith most likely had saved him from peril. Rushing into the house where Smith was sleeping, he kissed him on both cheeks, and pinned his medal on Smith's pajamas. The medal was the Médaille Militaire, the highest medal a French non-commissioned officer could receive during WWI, equivalent to the Distinguished Service Cross. Frank Smith gave Jim that very medal in 1938.

Jim would soon be on a mission, tracking down every WWI aviator he could, collecting their biographies, and other mementos, such as military uniforms long in mothballs. "The reason they let my father have the uniforms was that they knew he was sincere in his interest in preserving their history," said Andy Parks, on that day two years ago as he honored his father, nearby but only partially unable to comprehend all that was going on. Jim collected so many stories of WWI aviators that it was said that he could speak of the events that had occurred "almost as a contemporary." According to Andy, his father wasn't concerned about just collecting stories of American aviators. He believed that no matter what country they were fighting for, they felt that they were doing the right thing.

Jim Parks chose to practice medicine, specializing in obstetrics and gynecology.

He also earned a PhD, and set up practice in Aurora, Colorado. Dr. Parks became a pilot when he was in his late thirties. He instituted a WWI living history museum in his basement, where he kept memorabilia, and over 30 mannequins dressed in authentic WWI military aviator's uniforms, many of which were worn by famous Aces of both sides of the conflict, including German, Turkish, French, English, Austro-Hungarian and Italian. He also collected WWI aircraft parts and pieces, including propellers and aerial machine guns, and many other rare items.

In 1981, Dr. Parks co-chaired a reunion held in Paris for the surviving aces of WWI. Of 105, according to Andy, "Forty-five came for a round table discussion." Aces from America, Germany, Hungary, Austria and other countries came together to celebrate and trade stories of a shared experience from long ago.

In 1983, Dr. Parks organized and hosted the final reunion of the Lafayette Flying Corps, an American unit of fliers recruited by the French government. The reunion, which lasted several days, was held in several locations, including the Air Force Academy, Colorado Springs, and attended by surviving fliers from that unit, as well as other WWI fliers. The Lafayette Flying Corps made Dr. Parks honorary member number nine; Charles Lindbergh was number 8.

In 1984, Dr. Parks helped establish the WWI Aces and Aviator's Day, which was instituted by President Reagan in the White House, with eight American aces and their families present.

Dr. Parks founded and served as president of the Lafayette Foundation, a collection set up to preserve the history of the Lafayette Flying Corps. Eventually, the collection was expanded to contain other WWI aviation memorabilia, such as authentic uniforms, medals, commendations and equipment. It is currently on loan to the Aviation & Space Center of the Rockies/Wings Over The Rockies Museum, at the former Lowry Air Force Base. The collection at Wings holds uniforms and memorabilia of those pilots who visited the boarding house many years before, and others. The medal given to Dr. Parks in 1938 by Frank Smith is the first medal displayed in the collection today.

At Platte Valley Airpark, northeast of Denver, the foundation hangars replica WWI aircraft, including a Sopwith Pup and Fokker Dr.1 (Dreidecker), a tri-plane. The Fokker Dr.1 is similar to the one flown by Baron von Richthofen, "the Red Baron." The collection also includes "Aquarius," the Fokker D.VII built by Dr. Parks, and two S.E.5s.

Of six WWI replica aircraft, four are currently in operational flying status. An S.E.5 will be built, one half in fabric, and the other half in see-through material, so the curious can see how these planes are constructed. "We'll do a presentation with a balsa wood model, which will show kids you can build a real airplane," says Parks.

The Jeppesen Aviation Foundation donated money to the nonprofit organization for construction of the cut-away aircraft; Westwood College of Technology donated the manpower and expertise to construct the wings for the foundation's educational project.

When not in their hangars in Platte Valley, the aircraft can be found at various fly-ins and air shows. With his dad at his side, Andy put the display together at Wings. "By the time we were doing this, he was still quite capable of guiding me and helping me, but wasn't at the point where he could do what he would have liked," Andy explained.

At Wings, mannequins, many wearing medals, are styled to look like the pilots whose uniforms they wear, from facial structure to height and build. Actual logbooks and photos are in archives locked safely away. On display is the group of the original American pilots who were part of the Escadrille American, Americans who volunteered to fly for France in 1915, including Elliot Cowdin, the first American aviator to win the Médaille Militaire and the first of seven Americans to join the Escadrille.

When the Germans protested the American affiliation (the U.S. was still officially neutral at this time), they re-chose the name of the Lafayette Escadrille, for a French general who volunteered for the Continental American during the American Revolutionary War. The group adopted gold wings for themselves—the first American eagle wings - starting the tradition of wings for American aviators. The second set of pilots to arrive at the squadron received silver wings.

On the side of the Escadrille's planes was a "screaming" Sioux Indian. Long before political correctness came along, these pilots selected the Sioux as a symbol that they were Americans. Some controversy occasionally arises, as an emblem on the design's headdress resembles the Nazi swastika; however, to the Sioux, long before Hitler came along, it was a talisman of good fortune.

The various helmets, goggles and maps actually carried in the planes, including machine guns and sidearms are on display. The maps were drawn on wood to prevent them from flying around the windy, open cockpit. The collection includes memorabilia of WWI balloonists, as well as numerous trench art pieces, art created by someone in the ebb of battle, at the time of battle, or shortly thereafter, depicting their situation, from their perspective.

German pilots are also represented throughout the museum, which displays several Pour le Mérite awards, also known as the Blue Max, including the only award document known to exist, with original medal. The German collection includes uniforms, medals, original honor goblets (given to a German pilot after their first victory) and Sanke cards, the German version of our modern day baseball trading cards. German youth collected these postcards and sent them to pilots, who would autograph them and send them back. Included among the

Sanke cards are three of Baron Manfred von Richthofen, two signed by Richthofen and one signed by the famous ace's father. The card was in Richthofen's possession when he was killed, so Richthofen's father signed the card with a note apologizing to the boy who had sent it, hoping his autograph would be good enough.

When a U.S. Army officer was reassigned to the Air Service, they retained their original uniform from whichever service corps in which he initially trained. Thus, you have infantryman, artillery, cavalry and other uniforms with wings represented in the Air Service. Also, officers paid for their own uniforms, so the wealth of a particular officer could denote a better uniform; wings were hand sewn on each uniform, showing the differences from tailor to tailor. The better material in more expensive uniforms has helped in their preservation.

American pilots honored in the display include Bill Lambert, American's second highest scoring ace with 22 victories, next to Rickenbacker's 26. Lambert's ace status was almost overlooked. He didn't transfer to the American squadrons when the Americans entered the war, because he felt obligated to stay with the British who had trained him. On the back of a photo of the ace in the museum, Lambert wrote a note to James Parks, telling him that he just wanted to wear the uniform one more time before saying goodbye to it.

Pilots and their families continue to make tax-deductible donations of uniforms and memorabilia to the nonprofit foundation, in part to keep it together, but also knowing that the treasures will get the deserved respect.

Famous names appear in the display cases, including James Stetson of the Stetson Hat Company and Francis Lowry, an observer with the 91st Aero Squadron who was killed in combat. Lowry came from an influential family in Denver who worked to have Lowry Air Force Base named in his honor.

Pieces of fabric from actual aircraft include a fabric tear from the 94th Aero Squadron. What you won't see when you visit the museum is the huge room full of archived documents, logbooks, photos and other memorabilia that there just isn't room for in the museum display. However, historians use the room. Arrangements can be made in advance with Andy Parks. With a collection this extensive, the Smithsonian should have an interest, and they do. But Andy's not selling, because he fears that if placed there, there wouldn't be enough room to display everything. They've asked for certain pieces, but the Parks refused to break up the collection.

Memorabilia collectors such as Jerry Priddy, the owner of Richthofen Castle in Denver (a treasure trove of memorabilia in itself), considered Dr. Park's collection to be the world's largest personal collection of WWI aviation memorabilia. "He was the last of the primos," he said, after hearing of his friend's recent passing. Andy continues the collecting where his father left off, searching diligently for items they haven't yet been able to acquire, such as a Japanese WWI flying uniform, a Russian and a Belgium uniform. Families who have items, especially entire collections of noteworthy aviators, are urged to will them to the foundation. Andy Parks wants to stay with WWI as the limit of his collection for now, perhaps moving into the golden age of aviation through the thirties up until WWII, a period that he said is already well covered.

Behind every good man, there's a good woman; so has been the case with both these Parks men. Continuing the tradition of Dr. Parks' parents, Jane Parks, Andy's mother, entertained numerous houseguests; many WWI pilots would stay with the Parks family, sometimes for weeks at a time.

In the early days, the ever-increasing collection of memorabilia was kept in their basement, in display fashion. Andy recalls a time when a Time-Life photographer stayed with the family for three months, when they were doing a photo essay on the collection for "Knights of the Air," in Time-Life Books "Epic of Flight" series.

Jane Parks passed away six years ago, after suffering from Alzheimer's. Andy's wife Michelle also supports the vision. The ultimate dream of the couple is to combine the memorabilia and aircraft into one large hangar/living history museum. The museum would offer an audio tour, with much of what visitors hear in the actual pilots' voices, courtesy of a collection of 30+hours of audio.

A venture like this takes lots of resources. "We really need angel investors who don't want to see WWI aviation nostalgia go down the drain and can envision this museum," said Michelle Parks. One recent investor is aviation pioneer Harry Combs, a resident of Wickenburg, Arizona. He recently made a monetary donation to the Lafayette Foundation to remove and transport his hangar façade from his old Combs Aviation Corporation hangar in Denver at 38th and Dahlia; the present landowner donated the hangar facade. "Any remaining funds will be used towards a new hangar to house the Lafayette Foundation uniform and memorabilia collection and World War I replica aircraft collection," said Parks. "But we're still a long way off in funds to complete this project."

A Colorado Aviation Hall of Fame inductee, Dr. Parks has served as president of the Colorado Aviation Historical Society, and has been active in the Experimental Aircraft Association, Quiet Birdmen, the Daedalians, the American Society of Military Insignia Collectors, and a number of WWI aviation related organizations.

"My dad always had this love for WWI aviation and flying," said his son. "I followed in Dad's footprints. What fun we both had together." After developing Alzheimer's several years ago, Dr. Parks continued to attend many aviation functions with his son, often wearing one American uniform or another, a tradition Andy continues. Even two years ago, Andy was preparing for the day when his condition would gradually worsen. "It's been hard realizing that things are going to change," said Andy on that day at Platte Valley. But even then, Andy was taking steps to fill his father's shoes.

Andy Parks says that Dr. Parks and his wife Jane's children each inherited a different aspect from their father. "I followed the aviation side," he said. "There's an aspect of my father a lot of people don't know; he was quite an artist. Andrea, my twin, is an artist, like my father. Ted, our older brother, is a physician." As for his father, Andy says that he was a "great dad."

"He was not only a dad to his own kids, but to a lot of people," Andy said. "As an OB-GYN, he brought so many kids into this world. His patients thought of him as a fatherly-type person." Parks' passing came at a unique time. Andy and Michelle are expecting a child. "My dad is passing but a son is being born for the first time in this family," he said. "We know he is going to be a boy, and that he's going to have the best guardian angel a person could ever have."

Dr. Parks was honored in a memorial service (celebration of his life) at Wings over the Rockies on 13 September at 5:30 p.m. A fly-in tribute took place on 19 October 2002, at Platte Valley Airpark. Monetary donations can be made to The Lafayette Foundation, 1192 E. Michener Way, Highlands Ranch CO 80126. For those interested in donating WWI aviation memorabilia, please contact Andy Parks at (303) 995-2135. *The Lafayette Foundation, 1192 E. Michener Way, Highlands Ranch, CO 80126.*

Bibliography

Air Force Pamphlet 36-2241, Volume 1, 01 July 1999, Personnel, Promotion Fitness Examination (PFE) Study Guide, Chapter 3, Air Force History.

BOYNE, Walter J. *Beyond the Wild Blue, A History of the United States Air Force 1947-1997.* New York, St. Martin's Press, 1997.

CRAVEN, Wesley F., and CATE, James L. *The Army Air Forces in World War II, Vol. I: Plans and Early Operations, January 1939 to August 1942; Vol. II: Europe: Torch to Pointblank, August 1942 to December 1943; Vol. III: Europe: Argument to V-E Day, January 1944 to May 1945; Vol. IV: The Pacific: Guadalcanal to Saipan, August 1942 to July 1944; Vol. V: The Pacific: Matterhorn to Nagasaki, June 1944 to August 1945; Vol VI: Men and Planes; Vol. VII: Services Around the World.* (University of Chicago Press, 1948-1958). Reprinted for the Air Force History and Museums Program, 1983.

DONALD, David, General Editor. *(The Complete Encyclopedia of World Aircraft). The development and specifications of over 2500 civil and military aircraft from 1903 to the present day.* New York, Barnes & Noble Books, 1999.

FREEMAN, Roger A. *The Mighty Eighth, Units, Men and Machines (A History of the US 8th Army Air Force).* New York, Doubleday and Company, Inc., Garden City, 1970.

FUTRELL, Robert F. *The United States Air Force in Korea, 1950-1953.* Duell, Sloan and Pearce, revised 1983, 1991.

HALLION, Richard P. *Storm Over Iraq, Air power and the Gulf War.* Air Force Historian, 1992.

KAUFMAN, Daniel J., B.Sc., M.P.A., Ph.D. *Understanding International Relations* and *U.S. National Security Strategy for the 1990s.*

MAURER, Maurer, (General Editor). *The U.S. Air Service in World War 1. Vol. I: The Final Report of the Chief of Air Service AEF and a tactical History; Vol. II: Early Concepts of Military Aviation; Vol. III: The Battle of St. Mihel; Vol. IV: Postwar Review.* Air Force History and Museums Program, 1978, 1979.

NALTY, Bernard C. *Winged Shield, Winged Sword, A History of the United States Air Force*, Air Force History and Museums Program, USAF, Washington, D.C. 1997.

PARK, Edwards. *Fighters, The World's Great Aces and Their Planes.* New York, Barnes & Noble Books, 1990.

About the Author

Major Hal Skaarup is an Army officer presently serving in the Intelligence Branch of the Canadian Forces. Hal joined the Canadian Militia while attending the College of Trades and Technology in St. John's, Newfoundland in 1971. He later became an officer through the Reserve Officer University Training Program, and after completing his training at Shilo, Manitoba, was commissioned in 1973. In 1974 he completed his Bachelor of Fine Arts degree, graduating from the Nova Scotia College of Art and Design in Halifax. Between 1977 and 1981 he was a member of the Canadian Forces Parachute Team, the "Sky-hawks," participating in airshows across Canada and parts of United States. He is still an active skydiver with more than 1,700 jumps to date.

As a reserve intelligence officer, Hal served a two-year tour of duty in the Headquarters, Canadian Forces in Europe in Lahr, Germany from 1979 to 1983. In the summer of 1983, he transferred to the Regular Force and he was then sent to the Canadian Forces School of Intelligence and Security, first as a student, and later as an instructor. Between 1984 and 1986 he served as an intelligence analyst in Ottawa and then from 1986 to 1989 as the Regimental Intelligence Officer for the Canadian Airborne Regiment. As a Captain in the CAR, Hal went to the island of Cyprus on United Nation's duty from 1986-87. (In 1988, all those who had worn the "Blue Beret" on peacekeeping duty were awarded the Nobel Peace Prize.) The following year he completed the Canadian Forces Staff School course in Toronto. One year later he also completed the six month Canadian Land Forces Command and Staff Course (also known as Fox Hole U), in Kingston, Ontario.

On completing CLFCSC in 1989, Hal was again posted to Lahr, Germany where he served as the Intelligence officer for the 4th Canadian Mechanized Brigade Group, with the First Canadian Division (Forward). This came at one of the most interesting periods of change in modern European history. In 1992, he was posted back to CFSIS. On promotion to Major in 1993, he became the Officer Commanding the Intelligence Training Company, and later the Distance Learning Company at CFSIS. In 1994, Major Skaarup was posted to CFB Gagetown, News Brunswick, as the Intelligence Directing Staff officer in the Tactics School at the Combat Training Centre, where he was also "dual hatted" as the Base G2. Between instructing on and participating in numerous field exercises as the commander of the "Enemy Forces," he completed his Master of Arts Degree in War Studies through the Royal Military College, receiving his degree on graduation in May 1997 at Kingston, Ontario. In June 1997, he was posted to Sarajevo on a six month tour of duty as the Commanding Officer of the Canadian National Intelligence Centre in support of the Canadian Contingent of the NATO led Peace Stabilization Force in Bosnia-Herzegovina.

In June 1999, Major Skaarup completed the year long Land Forces Technical Staff Course with RMC's Department of Applied Military Science in Kingston. He is presently posted to Colorado where he is the Chief of the North American Aerospace Defense Command Exercise Intelligence Section, on Cheyenne Mountain in Colorado Springs. If you have questions or comments on the aircraft mentioned in the "Warbird Survivors" series, his e-mail address is **h.skaarup@worldnet.att.net**.

0-595-26223-6